A HIGH LOW TIDE

CRUX
THE GEORGIA SERIES IN LITERARY NONFICTION

John Griswold, series editor

SERIES ADVISORY BOARD
Dan Gunn
Pam Houston
Phillip Lopate
Dinty W. Moore
Lia Purpura
Patricia Smith
Ned Stuckey-French

A HIGH LOW TIDE

THE REVIVAL OF A SOUTHERN OYSTER

ANDRÉ JOSEPH GALLANT

The University of Georgia Press   *Athens*

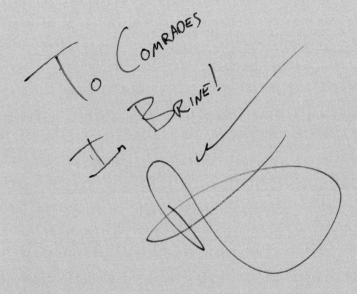

This publication is made possible in part through a grant from the Bradley Hale Fund for Southern Studies

© 2018 by the University of Georgia Press
Athens, Georgia 30602
www.ugapress.org
All rights reserved
Designed by Erin Kirk New
Set in 10 on 14 Miller Text
Printed and bound by Thomson-Shore
The paper in this book meets the guidelines for permanence and durability of the Committee on Production Guidelines for Book Longevity of the Council on Library Resources.

Most University of Georgia Press titles are available from popular e-book vendors.

Printed in the United States of America
22 21 20 19 18 C 5 4 3 2 1

Library of Congress Cataloging-in-Publication Data
Names: Gallant, André Joseph, 1979– author.
Title: A high low tide : the revival of a Southern oyster / André Joseph Gallant.
Other titles: Crux (Athens, Ga.)
Description: Athens, Georgia : University of Georgia Press, [2018] | Series: Crux, the Georgia series in literary nonfiction
Identifiers: LCCN 2018018102| ISBN 9780820354507 (hardcover : alk. paper) | ISBN 9780820354828 (ebook)
Subjects: LCSH: Oyster culture—Southern States. | Oystering—Southern States. | Oyster industry—Southern States. | Oysters—Southern States.
Classification: LCC SH365.A3 G35 2018 | DDC 333.95/5410975—dc23 LC record available at https://lccn.loc.gov/2018018102

CONTENTS

Preface vii

Introduction 1
Tidewash 9
The Spatking 21
Docents 33
Sunbury 41
The Spat Stick 55
Poachers 73
Out of the Water 83
Down at Charlie's 93
Joe and Lester 103
The Lone Wolf 117
Mud River 131
The McIntoshes 143
Spawn and Sputter 153
Earnest in the City 163
Mass Stimulation 175
Stumbles 185
Coastal Day at the Capitol 199
Clackers 211
Running the Rabbit 217
Oyster South 233

Acknowledgments 243

PREFACE

A High Low Tide is a work of nonfiction. The events reported within these pages took place between May 2013 and January 2017. Scenes depicted in this book were observed firsthand by the author. Descriptions of events not witnessed by the author were compiled from a combination of interviews, newspaper reports, historical archives, and public documents. All photographs were made by the author.

A HIGH LOW TIDE

A cluster of oysters on a mud bank near Wilmington Island, Chatham County, Georgia.

INTRODUCTION

On a crisp evening in January 2016, I sat at the bar of the Kimball House, a restaurant in the Atlanta suburb of Decatur. The interior was beautiful: mahogany-stained wood split the dining room into sanctuaries of hospitality; tall shelves stocked with spirits served as a cocktail reliquary; mirrors behind the bottles reflected light through green and brown glass; the wait staff glided between sound waves of conversation, silently efficient in delivery of entrees, their attention imbuing the loneliest bar patron with the confidence of a muck-a-muck.

I'm not the type of guy who often dines at fine establishments like the Kimball House, but I'm very much a practitioner of its gospel: oysters.

Perhaps most prized among bivalve mollusks is the eastern oyster, which science calls *Crassostrea virginica*. It is a small but intricate being, composed of heart, stomach, intestine, and other life-sustaining organs. Floating in brine flushed regularly through gills, an adductor muscle anchors the oyster safely between two shells connected by a hinge. Pop that axis and the salty protein is freed.

A most patient animal, the oyster is the hidden sentinel of our watersheds. It filters pollutants from estuaries as crabs nip at its shell and waits, like the Giving Tree, to be plucked in patronage to our appetites. Shucked and upturned, the oyster first slides a salty liquor down our gullet. In that water, place can be tasted—not just a splash of an Atlantic current, but of a specific bay or creek. The oyster spits the last salty gulp it took into our mouths. It is a field report, delivered with hints of iodine, melon, cut grass, or cucumber.

Eating an oyster transports us to the ocean. For a moment, we engage with a wild and wet frontier without need of boat or rain slicker. The connection bewitches us, and oyster bars are where we flock to be charmed. We are doing so in droves. Throughout the 2000s, oyster-focused restaurants like the Kimball House opened at a steady clip. Consumer trends drove that growth, but it started with a newfound availability of supply.

Throughout the twentieth century, natural North American oyster beds diminished from years of overfishing and human-driven environmental degradation. New England, New York, and the Chesapeake Bay all watched their oyster landings dry up. City edges bumped up against oyster habitats, polluting them to the point of catastrophe.

The advent of aquaculture, a scientific method of raising seafood as a farmer grows crops, promised to reverse course. Harvesting an oyster crop would require less reliance on nature and could be managed the way Nebraska farmers tend cornrows.

East and West Coast oyster farms tripled under the guidance of entrepreneurs, resulting in a twenty-first-century renaissance. In Virginia alone, a state that had struggled to maintain a healthy wild oyster population throughout the twentieth century, farmed oyster production increased 800 percent between 2006 and 2012. The bounty needed an outlet. Enter the oyster bar.

I walked into the Kimball House that night to inspect its spread of appellations, as the source location of an oyster is referred to by aficionados. I also planned to sate a craving. I had spent the morning watching oystermen from the Georgia coast speedily shuck their wares for legislators at the Georgia State Capitol. I tried a few but wanted more.

An ulterior motive also guided me to Decatur. I had hoped to talk aquaculture with Bryan Rackley, a partner in the Kimball House and something of an oyster sommelier. If oysters are a faith and the Kimball House one of its shrines, then Bryan is a bishop.

Bryan had helped start Oyster South, a small committee of restaurateurs and oyster scientists dedicated to raising the national profile of southern oysters, namely, bivalves from the Gulf of Mexico and the southeastern Atlantic Coast. He was something of a purist when it came to ordering oysters: he avoided large wholesale companies, preferring to deal directly with oyster growers. Like a chef walking pastures where grass-fed cattle graze, Bryan visited oyster farms whenever possible. He would head down to Dauphin Island, Alabama, or Caper's Island, South Carolina, inspect the operation, and drive home with coolers full of oysters in the bed of his truck.

Bryan went straight to the source for a reason. A problem with oyster aquaculture, he argued, was that it potentially marginalized traditional fisherman. Aquaculture isn't subsistence farming, and therefore it requires financial investment. Water-worn fishermen typically weren't swimming in capital. Keeping blue-collar fishermen solvent, Bryan said, was paramount. He did what he could to help by ordering seafood directly from oystermen, avoiding middlemen whenever possible.

When it came to oystermen, or watermen, as people who commercially harvest shellfish prefer to be called, I was specifically concerned with the small band that worked the coast of Georgia. I had spent the previous two years among the marshes of McIntosh, Liberty, Chatham, and Camden Counties. On the boats of watermen, I tried to understand the motivations, hopes, and dreams of the people who work daily on the water.

Georgia's coastline is roughly one hundred miles long, pegged at the north by the Savannah River and at the south by St. Mary's. A series of barrier and coastal islands mark its length and protect an essential and pristine habitat from Atlantic waves. Between mainland bluffs and island edges is a salt marsh, the second largest in the United States. A maze of rivers and creeks skirts islets of mud and sand.

There, fish, birds, and algae thrive. People too. Crabbers. Clammers. Oystermen. The marsh beckons them to reflective labor, where waning suns light butterscotch and chartreuse sea grasses.

Sadly, this majestic place went unrepresented on the Kimball House's oyster menu. I came to the Kimball House to brainstorm with Bryan about how this might change.

To prompt our discussion, we ordered oysters.

From a list of nearly two dozen varieties, we chose four: Daisy Bays from Prince Edward Island; Northern Crosses from Virginia; Humboldt Golds from California; and Caper's Blades from South Carolina. Soon after, they arrived, arranged along the edge of a stainless steel plate, a sprig of pickled sea bean snaked across the ice. A display of shellfish heraldry.

Our bar stools sat a four-hour drive from the freshest Georgia oysters, nearly two hours closer than Caper's Island, South Carolina. I did not need to ask Bryan why he sold Carolinian oysters and not our state's *virginicas* at his restaurant. I knew the answer.

Bad taste didn't keep them away. A wild Georgia oyster possessed spikes of saltiness enveloped in grass, an undeniably fresh and shocking flavor. But mess accompanied their brisk finish. The Georgia oyster's sharp, rugged shape and stubborn crags of mud, a by-product of estuary conditions, barred them from the elegant Kimball House.

"When we first opened, there were these two frat boys who used to go around Atlanta with a freezer truck selling seafood," Bryan told me. He is tall, with ginger hair and beard. Earnest in speech. "They had some from Ossabaw, if I remember correctly. We gave them a shot, but it just wasn't worth it."

They were a pain to shuck. Wild oysters amass in taloned bundles like roped chicken feet. Their razor edges are dangerous and undesirable, able to cut fingers and slice lips.

Too wild, too muddy, too much work. Lots of trouble and not enough payoff. Customers didn't care for wild ones. They looked odd and lacked consistent form. Farmed are the standard, Bryan explained: plump, clean, easy to open. That is what sells.

I held up the Caper's Blade, a thin oyster the length of a finger, less than two inches wide. Plucked from a marsh bed no more than sixty miles from the northernmost oyster lease in Georgia, this blade had passed a test oysters farther south couldn't. How?

"A lot of work," Bryan said. Both oysters possessed the same DNA and lived in the same rich estuary fostering such fervent oyster growth that the animals almost fought each other for survival. A clever waterman named Dave Belanger, "Clammer Dave" to his boosters, employed old-fashioned determination to craft a wild oyster worth selling to restaurants. He selected only the choicest shapes. He cleaned the crap out of them. He chipped away the gangly, brittle parts of their shells. Placed in protective bags, the oysters then went back into the water. Their shells healed into a groomed version of the original. After a few weeks of this restorative bath, the oysters were mud free and ready to roll.

Still, with all the work that went into Caper's Blades, they weren't always available. Bryan preferred farmers and was happy to support Clammer Dave. But lately, oysters from the Chesapeake and from Mobile Bay commanded the most space on his menus.

To join the iced ranks of Virginian bivalves at raw bars, Georgia oysters faced barriers to entry. Not insurmountable obstacles, but barriers nonetheless.

Many appellations vied for the same spots; UPS offered overnight delivery of farm-raised, clean, and perfectly sized oysters to restaurants, making Maine as close to Decatur as Savannah.

Dependable labor was hard to come by on the coast; factories and big cities lured away able bodies from the low pay and unreliable hours of oystering.

Bryan put it bluntly: to compete, Georgia oystermen would have to farm.

What I knew is that the transition from wild gleaning to modern farming had begun, albeit slowly. Three years earlier, a waterman, by force of will, had begun a campaign to transform Georgia's ugly oysters into belles. His story is the focus of this book. Contained within it are the personal, political, and cultural factors that stymied a local shuck at the Kimball House that night.

In the spring of 2013, I received a press release from the University of Georgia (UGA) that mentioned Justin Manley, a thirty-seven-year-old oysterman, one of ten in Georgia who still attempted shellfishing as a trade, and the youngest in the bunch by twenty years.

Justin and I spoke on the phone, and I could immediately sense he was more than a waterman. He had earned a master's degree in marine science, writing a thesis on oyster habitat restoration. While in school, he conducted aquaculture research for UGA. But he chose a career path outside the academy. Using idiosyncratic techniques developed through his graduate research, Justin farmed the first oysters in Georgia. His complicated method found modest success. He believed that with a little support from the university and state legislature, the industry could rebound. Time was of the essence, he argued, since competition was stiffening.

The South was primed to become the "Napa Valley of oysters," according to the prominent oyster writer Rowan Jacobsen. The Virginia Institute of Marine Science (VIMS) had come to dominate oyster hatchery research and promote the Chesapeake bivalve boom. Thanks to the efforts of Auburn University's marine program, dozens of oyster farms had come online along Alabama's tiny Gulf of Mexico coastline. A hatchery launched at the University of North Carolina at Wilmington helped oystermen not far up the coast from Justin start the first farms in the Carolinas. Justin didn't want Georgia left out.

Despite his marine biology background, Justin's push to promote local oysters had more to do with the livelihoods of his fellow watermen. Oyster stocks were healthy. They weren't at risk. But he worried that watermen

Justin Manley waist-deep in the Mud River, McIntosh County, Georgia.

and their way of life needed help. While chefs and journalists fussed over oysters as delicate vehicles of flavor to be placed on the same pedestal as heirloom tomatoes, the work of watermen was laborious, low paying, and unrecognized. He wanted to change that. Oystermen were also aging, and Justin hoped to build successful businesses for them before they retired or passed away.

Watermen deserved stability and proper credit. Together, maybe they and Justin could leave a legacy behind for the next generation. Beyond that, the product merited a spotlight. Justin wagered that the intense brine Georgia oysters offered would stand up against their rivals if given a chance.

Connecting the creeks of coastal Georgia with the global seafood distribution systems seemed a gigantic task. Foodstuffs whizzed overhead in airplanes. Eggplants and mackerels crossed international boundaries as daily practice. Could an ugly oyster from a state no longer associated with bivalves gain similar prominence? Justin assured me it could be done. But he didn't know that it would take more than the technical skill set he had adapted and implemented. Success couldn't be forced, he would learn. It would require adjustments, humbleness, and patience. An abandoning of dreams for practicality. A stepping out of the limelight. These attributes and abilities weren't necessarily natural to Justin and his beloved watermen. But the maturation and evolution of both, chronicled here, can, I believe, teach us plenty.

The story of a Georgia oyster rebirth was about more than aquaculture. Personality overshadowed science. Politics overcame passion. Relationships meant more than marine biology. In short, it was a story of people, not of an animal or an ecosystem.

If you research fishing history or hang out in fishing towns or with fishermen, one belief appears again and again, quoted in articles and books, and painted on faded life rings hung on shrimp shack walls: "It's no fish ye're buying—it's men's lives."

The English novelist Walter Scott composed the line for *The Antiquary*. On a walk, two noblemen encounter a fisherman's wife mending a net. They ask to buy some fish but protest at the price, offering a lower one. The woman sets her arms at her waist sternly and shouts them down. My husband and my sons go to sea, she argued at them, in horrible weather, and you think they should get nothing for their fish. She yells the oft-repeated statement at them. She is effective, and they meet in the middle.

Justin opened up his life to my questions and observations. He introduced me to the watermen whom he considered colleagues and heroes. He hoped I would chronicle the return of the oyster they had devoted their lives to. He offered to sell me a fish, but I wanted all the rest. The backstory. The drama. Desires dashed and scored. Oysters held less interest for me than the characters I met, Justin chief among them. This wasn't about oysters. It was men's lives.

TIDEWASH

Approaching a bed of oysters on the Crooked River.

Justin Manley thought he could make a late run out to Jones Hammock Creek. He kept his best oysters there: Sunbury Selects, the ones he had reared from larvae into beauties. Each one was molded by nature and nurtured by his attendance. Restaurants in Savannah, Georgia, paid a premium for those. And on a Friday, later than he would have liked, an order came in from his best account. Given the slim profit margins in the shellfish business, he couldn't leave it unfilled. He had a business to run. A wife, two children, to make proud.

Reports promised that storm clouds would come, but afternoon skies looked fair. The October sun had begun its western arc over pine peaks. He deemed it high enough to try a go. Waves lapped at the dock pilings where he had fastened his boat. He told himself they lacked aggression as a bracing wind gained pace and nipped at his goateed chin. There was still time to leave as the tide rolled out. He would slip back shore-ways on the incoming surge before thunderclaps potched the air.

East through the Medway River, he steered his sixteen-foot juniper-colored Crestliner. He called it the Green Hornet, *a small boat built for lake speeds and not prepped for slaps of salt water. Evergreens shrunk behind him as the trough of St. Catherine's Sound collected tributaries and pooled them before the prow. Justin pushed on, cheeks rosy and pecked by sea spray.*

He held the Green Hornet *to the inner channel. Standing at the boat's wheel, arms hung low to steer it with two fingers, he dwarfed the slim vessel with his muscular frame. In creeks like Jones Hammock, oysters grew plump. Deep tides drained the marsh dry twice a day. Low water limited access, since even experienced boaters couldn't float the narrow passages during shallow hours. Creeks served as vaults for shellfish, secured at the edges by tines of skinny cordgrass unfurling from the sulfur-scented muck.*

Justin cached market-ready product in marsh trenches, safe from predators and thieves.

He turned north into the creek and tilted his propeller out of the water. He could stretch out over the gunnels and, like a kid rattling a twig along a picket fence, run his fingers across spat sticks, a tool used in rearing juvenile oysters; he had adapted it from a French method. He hopped overboard into calf-deep water, grabbed a cleat-threaded rope, and hauled the boat onto a wild oyster mound. Shells crunched underneath its weight.

He piddled about his farm for a bit. He shook sacks of small oysters, checked knots in lines, and inspected the population for mortality. He heaved bags of oysters into the Green Hornet. They would be coming back to shore with him. Then, once cleaned, off to a restaurant.

Suddenly, his gut flipped. He looked southwest: a blue-black wall of fierce gales and rain was approaching with speed. The mass shrouded daylight's last brilliance. Time had trickled away; the storm sneaked in; the tide had not returned in kind. To beat the weather and the wicked waves it would bring, he had to leave now. But he was stuck, and he knew it.

Using his leg as a lead line, he measured too little navigable water. The still-barren creek bed barred his exit. Water would be no help in his escape; he would have to tow the boat himself.

Justin wrapped the bow rope around his fist like boxer's gauze and pulled the slack line tight over his shoulder, lifting the hull out of the mud with a heave. With one heavy lunge after another, he trudged out of the creek toward deep water. Each step took his full effort. The mud gripped his tennis shoes and slowed his march to a moonwalk. He waddled out of Jones Hammock as if his legs were plungers, compressing air between shoe soles and muddy river bottom, a suction reluctant to release.

The rope scraped his palms as the aluminum vessel trailed behind him. Through the slog, deep water in sight, nerves alerted his brain to a sensation that stress had kept passive. A wince. A laceration. Something singed the skin, a shock ran up the back of his legs. His pulse quickened, blood boiled. He knew the culprit.

Subtidal oysters, dagger-like buggers, cleaved slits into his shins. Salt water stung the wounds like a firebrand. He could do nothing to stop it. To escape, he had to reach deep river.

Justin bore the pain with the same stubbornness he had used throughout his life. Plow through, that was what he would do, no matter the obstacle, the catalogue of errors, the tweaks, and corrections. He was strong enough to do this. The fever behind his ear—an adrenaline poker—thrust him ahead. With one last gasp of strength, the waterline finally at his waist, he hoisted himself over the Hornet*'s gunnel. He gathered himself and plunged the propeller fast into enraged seas.*

After two cranks, the engine sputtered awake. The motor hummed and he knew then he was safe. The worst of it had passed. West into the Medway, the skiff fought savage waves and gave the tired captain a queasy ride. He piloted with white knuckles and wet clothes.

Justin survived. Scarred, but smarter.

Three years passed, and in 2014, Justin considered his luck improved. Educated by past mistakes, he had made drastic adjustments and expected only clear skies ahead. Nature would threaten him less, he figured, and any existential challenges that remained could be faced down with brainstorms instead of brute force. Anything was better than drowning.

But this was dubious optimism. For Justin, nothing would ever come easy.

On a tranquil June afternoon, not a dark cloud in the sky, Justin began to stir up conditions for a different sort of squall. He had been invited to help form this storm. He thought the disturbance he was planning would be relatively mild.

He waited outside a drab brick building on Skidaway Island, a back barrier island twenty miles south of Savannah, home to the University of Georgia's Shellfish Research Laboratory, part of the 700-acre Institute of Oceanography.

Two stories tall, the lab skirted a steep bluff by the Skidaway River. From its back door, a gangway stretched out over the water to reach a dock with sailboats, skiffs, and cruisers roped to posts. From the front door, a road that connected the campus to the rest of the island cut through a clearing between river and forest, where treetops cast elongating shadows across trimmed Bermuda grass as a balmy breeze mumbled through pine needles and oak leaves.

Dawn rising over the Crooked River.

Justin and I leaned on a white truck marked with a university logo, waiting for his guests to arrive. A meeting of watermen had been called that night, one that would mark a turning point in Justin's life. Built like a linebacker, he tried to mask nervousness behind sporty prescription sunglasses and an ever-present Detroit Tigers baseball cap.

He had come to terms with a new truth: He could no longer call himself an oysterman. Tonight, he would introduce his former colleagues to a new Justin, an extension agent, a university employee, an aide to shellfishermen. We discussed the anticipation hanging heavy on his mind while the summer air weighted us with sweat. He was eager to start fresh. He knotted his T-shirt in his fists, as ready to prove himself as an anxious racer backing into the starting block.

Since that elemental battle in Jones Hammock Creek, Justin had learned lessons. The first: capital and other obstacles would bar him from working the water. The fifth: maybe his path was scientific, not entrepreneurial. Lessons two, three, and four were harder to talk about.

He had made decisions, difficult ones, to ward off past chaos. He had dry-docked the *Green Hornet* and, with it, his dreams. Easier ways to make a living than oystering in Georgia, especially the way he went about it. He had refused to harvest wild oysters. He couldn't be bothered with clams. He had obsessed over farming despite regulations, cultural opposition, and the water's constant hazards. An oysterman's life challenged him daily, and his peculiar, inflexible approach hadn't made it easier.

Other watermen lived as lone wolves. They denned close to the marsh, in tandem with it, and trimmed off any excess in expenditure. They offered relationships only to kin, kept the wide world obscured by avoiding it.

Justin was different. He made his home in Savannah, a big city miles from rural docks. He had his family, whom he loved more than anything. Mortgages. Health insurance. Tuition. Vacations. Retirement. Happiness. He weighed oysters on one scale, wife and kids on the other, and the latter sunk. The choice became clear.

He quit oysters. He told himself he did so for his family, that they needed a more dependable father, someone who could provide—the directive that brutish men use to shackle themselves and their families in tradition. They hadn't asked him to, but he couldn't be convinced otherwise, despite protest from his wife, Amelia, equally as tall and more agile than her waterman husband. He heard none of it. He had made a choice.

On an early June evening, none of the ramifications of that decision mattered. That, in the name of making a living, he had left the marsh for a toxic work environment. That Amelia watched him fall under a depressive shroud. That he went full curmudgeon. Retreated, went dark, incommunicado. For his family, it was a turn far worse than not providing.

He nearly drowned under false expectation. This evening, he shrugged off the past and readied himself for new responsibilities. Finally, with extension tasks ahead, he could think of something other than himself. He had a job to do: improve the working lives of Georgia oystermen and the reputation of wild oysters alongside them, and do so with aquaculture, modern techniques never before incorporated in Georgia. He was perfect for the gig. As Justin saw it, his failures had expertly schooled him for this specific assignment. All that was left was to prove it. The opportunity to do so began as trucks pulled into the laboratory parking lot.

Here came the watermen.

Oystermen rarely came to the shellfish lab. Extension agents, if need be, visited them. The lab was less a full-tilt science factory than a base where researchers computerized data collected from marsh experiments. Neither activity gave shellfishermen a good reason to pay the lab a visit. Their leases were miles away, and leases mattered most. But this night, watermen were making an exception.

Justin considered watermen to be a second family. They were big brothers whom he looked up to and hoped to emulate. But just like any family, the rifts between members could be raw and unforgiving. Some blood relatives could hardly stand the sight of one another. Justin believed he could unify them all by dangling a carrot. Let me help you make more money, he offered. Intrigued, watermen were coming to Skidaway to find out more.

"They don't make much money, but you can't tell by the spirit of how they live," Justin told me. "They work so hard in the most extreme conditions all day, and they still make it work. They're proud of what they do."

Trucks rumbled into the lab lot. Their license plates told of how far they had ridden: Liberty, McIntosh, Glynn, Camden. As close as forty-five minutes, as far away as two hours, they left behind rich estuaries protected from Atlantic waves by barrier islands: Little Tybee, Wassaw, Ossabaw, St. Catherines, Sapelo, and Cumberland.

They creaked open cab doors and stretched their legs, and reintroductions began.

Dick Roberts of Whitehouse Seafood in Woodbine and Dan DeGuire, his right-hand man, arrived first. They worked waters near the city of St. Mary's, the last stop before the Florida border, to stock a retail store. A third partner in the business didn't make the trip: Dick's wife, Karen, had stayed home. Plenty of Camden County locals would be stopping by for mullet and shrimp, and someone had to mind the shop.

Dick's arms shot far beyond the ends of his long-sleeve shirt, the stitched cuffs of his pants exposed his ankles. His clothes looked too small for his tall body, a man with fish, not fashion, on his mind. A wool Greek fisherman's cap, though, fit perfectly on his head, his still sandy hair tucked underneath. The hat never left its perch. I couldn't help but notice his hands, big like a sourdough boule and fingers like breadsticks.

Dan had dressed up for the occasion. He wore a white collared shirt tucked into khakis, rainbow suspenders, and a bowtie. A short, balding French Canadian, Dan carried a binder full of photos taken around the shellfish lease he worked for Dick. Dan loved talking shop in detail.

Dick, Dan, Justin, and I leaned on Dick's truck bed, talked about how things were going down in Camden.

"Well, this year we had no wild spat collection. None," Dan said. Oysters spawn a few times a year, replenishing natural beds, but that hadn't happened, Dan said. Not too long ago, their lease experienced massive die-offs, but had rebounded. The waterman knew it was, perhaps, just a natural phase, but the bad set troubled him.

Justin listened, nodding. He hadn't seen some of these guys in two years, maybe more, and the chance to hear oyster talk from honest-to-god watermen invigorated him. That Dan described a disturbing occurrence in the ecosystem proved a strong selling point for the introduction of aquaculture in Georgia. This was a problem Justin could solve. The eagerness to do so made him jittery.

Justin motioned for us to head inside the lab, into a classroom where chairs had been set in rows, where we would wait for everyone else.

Justin's new boss, Tom Bliss, joined us there. An expert grant writer, Tom had long hair that raced down the back of a polo shirt. Small in stature, his shorts barely covered the muscular legs of a distance runner. For everything that would unfold from this evening, Tom had coordinated the logistics. Justin would enact it.

Joe Maley from nearby Liberty County arrived next. Joe served as current president of the Georgia Shellfish Growers Association (GSGA), a loose industry group to which most watermen belonged.

Justin braced his hips against a concrete half-wall. Arms crossed, a library of brochures about coastal restoration and shellfish-handling safety fanned out behind him in front of a bank of glass.

A man dressed in a brown vented fly-fishing shirt and angler pants joined him against the sill. He stood just as tall as Justin but looked a decade older. Gray streaks sliced through close-cropped black hair. He smiled at the shellfishermen now finding seats. Stitching on the chest of his shirt explained his presence: Dom Guadagnoli, Georgia Department of Natural Resources, Coastal Resources Division. Dom was head of shellfish for the division, the guy who guided and permitted the livelihoods of every fisherman in the room.

Jeff Erickson lumbered in. Jeff had grown up on the marsh, fished since he could walk; he still shrimped, clammed, and oystered. Jeff was in his fifties, and though the sun had kept his hair bleach blond, the salty air had weathered his skin.

Down in Eulonia, Jeff lived next door to his brother, Mike Townsend, an equally accomplished waterman. The siblings hardly spoke. The brothers had little use for dialogue. Mike chose to stay home. The decision surprised no one, as the presence of the man who walked into the classroom next—Charlie Phillips—would keep Mike away from his own funeral.

Gossip rarely sounded within marsh boundaries, lost amid the hypnotic caws of shore birds. Speaking ill was saved for regulators. But watermen discussed one feud without fear, without whispering, and without coming off as gabby: Mike Townsend did not enjoy the company of Charlie Phillips. It was a statement as trustworthy as the tides, but trying to find the origin of the vitriol led only to conjecture.

"I just don't trust him," Mike liked to say when he was being kind.

Other watermen told me they exercised caution when dealing with Charlie. He was the apex predator in the shellfish game in these parts; many clammers and oystermen sold to Charlie. Nobody boycotted him, but they proceeded with caution. Why keep money out of our pockets? they said. Just make sure it's all there.

Except among shellfishermen, Charlie was a beloved figure.

Often introduced with the title "captain," Charlie had a reputation for voraciously supporting environmental causes, for raising his voice in protection of marshes and rivers. He was cunning and smart, delivering no-nonsense one-liners about why wetlands mattered and throwing his economic weight behind causes that worked to keep that ecosystem healthy and pristine. When he delivered his quips at charity functions, they dripped brine like an anchor hauled out of the water. Charlie clocked plenty of city miles with his marsh bona fides.

Charlie built his business—Sapelo Sea Farms—first with shrimp boats and fishing vessels, then with farmed clams. Raised from seed imported from Florida, Charlie's littleneck clams wound up in restaurants and grocers as far away as Toronto, Canada.

His success wasn't without critics. Some felt jealous of the market share Charlie had cornered with Sapelo Sea Farms, felt envious of his public stature.

"Charlie is the man around here," Jeff once told me. "The rest of us are just chump change."

Why Charlie was such an antihero was hard to discern. Nobody offered any specific proof of malpractice, at least not with outsiders around. No visible protests erupted when he walked into the GSGA meeting, a calm grin beneath a bushy gray mustache, and slouched down into a rear chair.

Two men, father and son, entered just as quietly. The McIntoshes, Earnest Jr. and Sr., descendants of enslaved people who still carried the name of one of Georgia's founding settlers. Alongside Jeff Erickson and Mike Townsend, the Macintoshes could boast of the longest-standing relationship to the region. Decades earlier, Sr.'s father had run a successful crab plant that employed dozens in their hamlet of Harris Neck, a historic and contentious African American enclave at the northern edge of McIntosh County. If Charlie's presence gave the meeting business clout, the Macintoshes added authenticity to the proceedings. They did so with grace: their lanky bodies sank into seats, and Sr.'s eyes, green like an aged emerald, calmly surveyed the crowd without comment.

Another waterman was notably absent. John Pelli, the only shellfisherman in relatively urban Chatham County, couldn't make it, Justin said later. John lived not too far from the lab, and Justin could watch John's boat speed through the Skidaway River, past the shellfish lab, on the way to his lease in Wassaw Sound. John held the northernmost one in the state, closest to the biggest market for shellfish in Savannah. An upscale grocery store stocked John's clams. If oysters returned to prominence, John already had connections to sell them.

There was one last attendee on this evening. He was the most recent addition to the bunch: Rafe Rivers, an organic farmer newly relocated to the coast. A young entrepreneur with an interest in oysters, Rafe hadn't yet attempted shellfishing, but he could smell the market for prime half-shell oysters. A ponytail streamed down a flannel western shirt that Rafe had tucked into Wrangler jeans. A brass belt buckle tied the outfit together. He stood against the back wall, made eye contact with Justin, spoke to nobody, and looked very serious, perhaps intimidated.

Justin rose to address the group. His heroes seated before him, the resilient watermen whom he had hoped to emulate. Oystering aroused great pride in him, yet the tension of the job nearly choked him. In his eyes, he had never matched the reputations in the room. But he had been given a chance to help them.

Everyone mustered this night would play a role in returning the Georgia oyster industry to past glory. Success would be built on these relationships, by crossing fissures, repairing mistrust, ensuring each man's skiff rose equally on the new tide. That was the plan, anyway. And it would start tonight, with Justin central to its progression.

Justin readied a bass and professional voice in his throat, the surety of the timbre unfamiliar to him. Some of you remember me, he said; some of you haven't seen me in a while. Back then, they knew him as the Spatking. A shellfisherman with a novelty item—the first farmed single oysters from Georgia. He had worked the water as a lonely visionary.

Now, much had changed. No map charted what would come next.

THE SPATKING

Atlantic waves breaking along a beach.

Atlantic waves kissed his toddler toes, sun rays warmed his pale skin. He poked fingers into white quartz crumbles and glopped plastic shovelfuls of Dayton Beach sand into a bucket. Parents looked on from beach chairs, his little sister gestating in his mother's belly, as a dehydrating breeze blew over them and whistled through the grid of vacation homes beyond development's beachhead.

Warm water splashed him and washed away his sandcastles. He rebuilt each one the water toppled, and he felt intense peace. He swore he did.

Could a three-year-old read his fortune in a tide pool? Justin believed it so. Haze affixed itself to much of his memory, like honey on a canning ring, but on this one thing he felt clearheaded. Forty years later, the details of an afternoon on Daytona Beach, Florida, hadn't lost their clarity. The moment the ocean lured him. The moment that determined his future. The moment when a boy born near Lakes Erie and St. Clair chose warm salt water over frozen fresh. The moment that would kick off an obsession with deep waters and filter feeders rather than the walleyes of inland lakes.

Justin had one problem. He lived in Michigan. His fate included cars, not boats, not marine life.

Behind a chain-link fence in the Detroit suburb of Chesterfield, Justin grew up destined to follow the school-to-automotive-factory pipeline laid for him by his father and grandfather. His mother taught preschool once her kids were grown, but had stayed home to rear Justin and his sister while her husband made a good living at a Ford Motor Company plant. Justin doesn't call this middle class, at least not anymore. It used to be middle class, he said, but it was just how auto industry people lived.

"It was a good childhood," he remembered. "I mowed lawns for five dollars a pop. My friends and I played could go play in the woods until they chopped it all down to make more suburbs."

His grandfather had moved north from Tennessee to join the postwar automotive boom. He had been a farmer, but couldn't resist a call from a brother who lured him with promises of high wages and safety from agriculture's finicky seasons. The elder Manley quickly hired on at a Chrysler plant. Factory life suited Justin's grandfather, and the dependable comfort kept the next generation on the line.

Justin finished high school, where his broad build found success as a football player, and enrolled in both Central Michigan University and the factory. He went Ford. For eight years, beginning in 1995, he cycled through a series of assembly-line tasks—which is what most workers do, he said, and didn't signify a lack of skill or interest—but mostly built rear seats for Town Cars and Mustangs.

Justin the laconic assembly-line worker had not yet sampled his first oyster. Just as he kept the ocean's first enticing touch well logged, he could never forget the first shuck and slurp. He was young, in his early twenties. On spring break in Panama City, Florida, capital of the Redneck Riviera, a beach populated almost entirely by snowbirds like himself. He went to a bar with friends, got drunk, and tried hitting on a woman. She was daring dudes to down raw oysters. Everyone shirked the challenge. Except Justin. He was smitten.

"She was hot. Don't tell my wife," Justin said, digging into fifteen-year-old memories. His chest chuffed and heaved with laughter.

He took down the oyster, an icy rush warmed by a drop of hot sauce. He returned to Detroit without the woman's name or number. His love fell elsewhere. He decided to pursue it doggedly.

While Justin channeled energy to the factory and the classroom, his mind thought only of the sea. He bided his hours, days, weeks, and years at the plant by filling his head with biology. He completed his studies in 2003. Immediately, he quit Ford, a separation that workers refer to as "processing out."

Justin had spent his last years in Detroit obsessing over the consumption of oysters, tasting offerings from across the country: Wellfleets and bluepoints from the East, and Kumamotos from the West. Urban Detroit imported enough bivalves to keep Justin's interest in oysters alive as he processed through and out of Ford. He prowled raw bars, eager to chomp on new varieties from a body of water he had never heard of before, much less swum in. I could imagine Justin at a fish market, fawning over oysters packed in chipped ice, as eager as a famished grizzly bear tracking salmon in the upstream run.

Justin soon directed his hunt south. His sister had relocated to Savannah for work, so Justin included graduate marine science programs in the area as potential sites of further study. On one such research trip, her apartment became his home base while he surveyed Savannah State University. She owned a condo on Wilmington Island. One night, he walked down the road to a bridge that crosses Turner Creek. He knew better than to eat oysters found close to sewage systems, but he wanted a closer look. He had never seen oysters in the (relative) wild. The sight of oysters in their natural habitat—an expansive reef the likes of which he had never seen—made Justin's brain thrum like a Tesla coil.

Weaponized by curiosity, he moved his investigation closer. He jumped the dock railing and landed on the oyster mound. With no experience in traversing the quicksand-like surface, he lost his footing. His shoe sank into the mud and he buckled. His knees bent forward, and his shoulders flopped backward as if he had been tackled. His arms shot out to brace his body for the fall, and his right palm met the brittle tip of an oyster. The point sliced his hand and drew blood. So much blood. He worried someone would see him and think he had committed a murder.

He winced when he saw the wound, then shook his head over how foolish he had been. A feeling the opposite of pain overcame him. He felt euphoria, again, the same comforting wash he had experienced as a toddler. This was Justin at twenty-eight: a sharp chin bare of goatee, hair covering scalp areas that are now bare, shorts splotched gray from sitting in the mud, forearms perched on knees, small and focused eyes fixated on the mélange of plasma and mud swirled on his palm.

An oyster injured him, sure, but he also believed it contained an oracle amid flecks of shell and blood droplets. Justin's father had instilled in him a sincere connection with nature, that people are part of the land and sea. This injury, he thought, was the transcendent moment when nature revealed a truth to him.

"Your brother will rise to meet you," he told me. "That's what I believe. It will show you how you are connected. Here I am being stabbed in the hand by an oyster, and I feel nothing but joy. I immediately knew we were part of each other. To this day, I'm able to connect with that animal."

Justin expected a scientific relationship with oysters. He would learn to respect the animal's adaptability, how it used the natural environment, took from the water only what it needed to survive. Of course, he found oysters especially delicious. As he recovered from that night, weighing its context again and again, he couldn't deny the declarative power of the event. More than a career, more than a field of study, oysters would be his life.

"It was our true introduction," Justin said of the incident.

A bloody handshake? I asked him.

"We're blood brothers."

Justin knew Georgia's estuaries grew abundant oysters. He couldn't understand why nobody took advantage of that bounty. Oysters seemed a treasure that few cherished. Seafood was an integral part of life on the coast, especially during celebrations. Folks loved their shrimp. But in some ways, Justin realized, the commitment was a façade.

As a grad student at Savannah State University, he attended his first southern-style oyster roast on the Isle of Hope, a posh community near Skidaway. This tiny Savannah suburb held barely 2,000 residents on less than a square mile of land. Historically, thousands of oysters had grown along the sides of the Skidaway River, which marked the isle's southern boundary. Guale natives survived winter months by harvesting shellfish in the estuary. But the number of oysters living and breeding and eating in waters off the Isle of Hope had long since diminished. They were out there, mud covered and hidden among the stalks of tall grasses. If Justin followed the sturdy grass-topped land until its edge was blurred by mud and then waded into the goo, he would find some.

Justin watched roasting oysters get shoveled off a metal grate and then dumped onto a newspaper-lined table. A horde of hungry people shucked them with plastic-handled knives, flicked the meat onto saltine crackers, and doused them with hot sauce.

The oysters roasted that day were sourced farther from the Isle of Hope than Justin could imagine. He asked the roast's coordinators where the oysters came from: Apalachicola, Florida. He was visibly shocked. I could imagine him leaning over the person, his frame casting an intimidating shadow, his midwestern upbringing disavowing the southern tradition of keeping one's mouth shut, and lecturing the fellow on the perfect salinity of the coastal waters here. It's an exceptional oyster habitat, he might have said. Don't you know that?

Here is what happened.

For a brief time between the decline of mid-Atlantic oyster stocks and the rise of Gulf of Mexico bivalves, Georgia dominated the early-twentieth-century oyster industry in the United States. Savannah, the jewel of Georgia's coast, added bivalves to the list of seafood its companies processed and shipped to inland accounts. In newspaper advertisements, the historic city's restaurants boasted sales of Belvidere oysters, pulled fresh everyday from the oyster beds south of the city, at twenty-five cents per quart. Sales were brisk. Back then, a man named Armenius Oemler was oystering's biggest proponent. In 1889, he published a paper, "The Past, Present, and Future of the Oyster Industry of Georgia," that argued for conservation of the state's impressive oyster stocks in order to prevent overfishing, as had happened in New York. Although Oemler ushered in an era of oyster conservation, he failed to build a successful business from shellfish. But an Italian immigrant did.

In the early 1900s, L. P. Maggioni and Company began to expand and control the oyster industry in Georgia and South Carolina. After the invention of canning equipment in the late 1800s, the business found a use for ugly oysters. They could be harvested, shucked, and sealed in cans, then shipped anywhere in the world. The Maggionis proved to be expert marketers of the product. The dozens of canneries that opened under their banner employed thousands of Georgians from Savannah to St. Marys.

The oyster boom the Maggionis led continued for decades. But the bust would come. The company complained that development along the coast ruined oyster health. And the advent of minimum-wage laws whittled down their cheap workforce. Bosses could no longer offer piecemeal pay to the poor African Americans who toiled in shucking houses and on harvester sloops.

A memorandum sent in 1956 to the board of directors of L. P. Maggioni and Co. laid out the remaining threats to the industry: "Foreign oysters are being imported in increasing quantities with damaging effects on domestic sales, and increased sewage and pollution actually threatens the end of the industry in some areas." In 1991, the Maggioni company, a shellfish powerbroker for 121 years, finally called it a day. The Maggionis sold their signature can label—Daufuskie oysters, recognizable by its native chief in full headdress logo—to a Korean company. Shrimp became the signature Georgia catch.

Only the sack trade in wild oysters remained. Except among a few loyalists, demand for them dried up.

"We don't carry Georgia oysters anymore," a Brunswick, Georgia, seafood distributor told an *Atlanta Journal-Constitution* reporter in 1981. Florida's Apalachicola oysters had begun their ascent, and customers already preferred the stony mass and meat of Gulf selects.

The distributor blamed state regulations for keeping a sizable oyster crop from coming to market. In Georgia, shellfishermen couldn't use a dredge—a steel scoop dragged by boat along the water bottom—and the state allows harvesting only in the intertidal zone, meaning the stretch of open riverbank exposed to fresh air between high and low tide. Big, fat oysters were left underwater and unharvested because of these rules, he contended. Scientists couldn't prove the claim, but tended to disagree.

Oystermen employing dredges wouldn't have saved the industry. In addition to the labor laws and global economic disruptions that upended the Maggionis, oysters just died. In the 1980s, a lethal combination of drought, elevated salinity, and parasitic protozoan infections ravaged Georgia's oyster population. Harvests dwindled. Many oystermen found new occupations. Dedicated oyster eaters sourced theirs from the Gulf.

A small number of patrons kept a handful of harvesters in business. When I visited Russo's Seafood in downtown Savannah in 2015, one of the old-school fish markets started by Italian immigrants generations earlier, local oysters were a specialty item, seldom in stock. Fat Virginia and Gulf oysters sold well, the owners told me. Around the winter holidays, when family dinners traditionally called for oyster stuffing, pints of shucked wild oysters from South Carolina became popular.

In the less dense coastal towns between Savannah and Jacksonville, a few wholesalers supplied local oysters. But there was always a demand problem. A distributor couldn't call up a Georgia oysterman, order a bunch of bushels, and expect him to deliver with any speed. Trucking some in proved far easier.

Oysters became speakeasy hooch, submerged under tides, barricaded by shoals. To source them, you had to know somebody: a waterman with a stash hidden deep in the marsh.

Justin spent his college time in Savannah at the University of Georgia's Shellfish Research Laboratory. He studied under Randy Walker, who had helped establish the laboratory back in the 1970s. Back then, Randy desperately wanted to revive the oyster industry. But by the time the mortar on the shellfish lab had dried, oystermen had opted for obscurity rather than developing their trade. New tricks held no promise. Randy's bosses told him not to bother with those old-timers. Instead, try clams. There was less history, less baggage. Maybe you will convince shrimpers to give them a shot. He followed orders and found fast success when he introduced clam farming in the 1990s. By the time Justin began spending time around the shellfish lab, clam sales had surpassed a million dollars.

But Randy couldn't ever really leave oysters alone. In making clams valuable, he had hoped to clear a path for their return. As a student, Justin would try to push away a few decades-old obstacles.

During his tenure, Randy conducted studies and collected data—the kind of currency that commercial, academic, and governmental partnerships are based on—and prepped for future oyster success.

He monitored the overproduction of oyster spat in the marsh, which had to be tracked, Randy told me, to prove that estuaries in the southeastern Atlantic were unlike those in the northeastern Atlantic, the Pacific Northwest, or the Gulf Coast. He recalled attending shellfish conferences in the 1970s and 1980s. When he relayed stories about the fecundity of Georgia's oyster population, his peers laughed. Impossible, they said. It took years, but he showed them.

Randy launched a final series of projects in the 2000s, titled "Overcoming Constraints to the Development of a Sustainable Eastern Oyster, *Crassostrea virginica*, Aquaculture Industry in Georgia." Justin helped Randy run experiments.

Oyster farming required oyster seed, which meant collecting wild larvae from open water and rearing them into adults. To do so, Randy and Justin had to catch oyster progeny, called spat, at their most microscopic size,

Workers for the Savannah Clam Company harvesting oysters.

two to three weeks after egg had met sperm. The idea was to mimic wild settlement, when oysters attached themselves to hard subtidal surfaces.

Justin tested different media for rounding up spat—bamboo, concrete, roofing tiles, and pvc. He set them out deep in the marsh and watched what happened. He saw how spat overset and biofouling—like the presence of barnacles—varied around the marsh, and learned where along a river's course he could obtain the healthiest and largest number of spat. If he placed growing oysters up tidal creeks, Justin determined, he could ensure their survival and an unblemished product.

Back in the lab, he counted oysters affixed to various media. His eyes totaled the little ovals, flat and an eighth-inch wide, with speed, faster than other researchers. His calculations earned him a nickname: the Spatking. He accepted the sobriquet. He enjoyed the humor of it. The name stuck.

The foundation for a career in oysters began to solidify on Skidaway. On weekends, he found time to round out his life with romance.

One night, at a bar in City Market in downtown Savannah, Justin's sister introduced him to a woman. With blond curls that framed a sharp jaw, and legs far longer than his, Amelia wowed Justin. Loud and opinionated, a physical therapist with job offers aplenty, she spoke her mind, with a hard Ontario accent and without apology. She gave Justin no evidence she might be interested in him. Nervously, he mustered the confidence to ask for her number anyway.

Amelia found plenty to like about Justin. He was masculine, athletic, shy, and kind. He was driven, committed to his work. His entrepreneurial dreams, as yet unrealized, excited her. They courted, and soon Amelia became pregnant. They married in Savannah, and Amelia gave birth to their first child, Bella, in the city. Justin concluded his research with Randy and accepted a university job in Oregon. If he was going to pursue marine science at the academic level, it was the starter position he needed. Amelia traveled back to Ontario with their daughter, but the United States denied her reentry, so Justin gave up his position at Oregon State University. Such distance was no way to begin fatherhood. Family would always come first. He moved to Canada and found work farming rainbow trout. Mercer, their son, arrived next. All the while, Justin waited for an opportunity to oyster. He wanted to implement everything he had developed under Randy's tutelage. Three years after Amelia's visa refusal, the family of four returned to Savannah.

In 2011, a lease came open in Liberty County, nearly an hour south of Savannah. Justin seized the opportunity. He already owned a boat, the *Green Hornet*, which he used to fish walleye back in Michigan. He purchased gear to gather spat and grow oysters to proper size; he wouldn't be bothering with wild oysters; he would capitalize on the dearth of beautiful single oysters; he would corner the market.

Months later, Justin strolled up the back doors of Savannah restaurants, carrying bushel bags full of his Sunbury Selects on ice. He shucked a few samples and made easy sells.

The Spatking had arrived.

DOCENTS

A shucked wild oyster.

Nothing about farming oysters came easy. The first crop of Spatking oysters sold out, but profit margins were slim to zero. Savannah's local single-oyster scarcity helped Justin's Sunbury Selects attract media attention. He had given the ugly oysters a makeover, and reporters and photographers flocked to see how he did it. He became a docent of the marsh, guiding newspaper columnists and public-radio reporters on river tours, passing oystercatchers, crab pots, and spartina (cordgrass), like paintings on a museum wall. Oysters were a long-forgotten acquisition that Justin found tucked away in a basement vault, dusty and unloved.

He looked the perfect tour guide, often dressed in green rubber bib overalls. Journalists' microphones captured the purr of Justin's outboard motor as they accompanied the oysterman on his thirty-minute commute from dock to lease. Over the din, he pointed out the barrier islands.

That's Ossabaw to the North, he said. Over there, to the South, is St. Catherines. Shoot out between the two, he gestured with a two-fingered thrust, and you've got the Atlantic Ocean.

Dolphin fins broke the waterline as a sea breeze kissed the journalist's cheeks. Pelicans barreled overhead. The marsh sold itself: the oysters, a bonus.

Brittle bivalve shells cracked under their footsteps. The liquor and meat of Justin's oysters tasted like the sour hint of marsh plants. Journalists marveled at his methods, how he transformed gnarly-looking wild oysters into a product worthy of high-end restaurants.

Rightly, they noted the price of Justin's oysters: four to five times as expensive as wild ones. He didn't admit to them that he was hardly breaking even.

Justin decided to take his aquaculture show on the road. In January 2013, he planned to attend Coastal Day at the Capitol, a lobbying event at the Georgia State Capitol organized by the Department of Natural

Resources. He figured he would shuck his prized farmed oysters for legislators and interest them in supporting the burgeoning industry.

A few days before the event, Justin drove to an empty lot, hardly big enough for a house, in Thunderbolt, a suburb of Savannah, where he kept equipment and a large commercial cooler.

A century ago, shrimping and fish-processing plants kept Thunderbolt's economy chugging along. By the 2000s, it had evolved into a middle- and working-class bedroom community. Thunderbolt's civic boundaries blended with Savannah's other southside suburbs, making the drive from historic downtown to marsh edge a visually seamless one. Marine shops still supply the yachts and hobby fishing boats now berthed at Thunderbolt's marinas. But on a slim plot, about an eighth of an acre, barricaded by molding clapboard tear-downs to one side and plastered ranches to the other, Justin kept Thunderbolt's commercial fishing traditions going. Which meant fishiness had returned, since he empowered eggy marsh scents to loiter near open windows in ways they hadn't in decades.

At the time, the Manleys lived in an apartment on Whitemarsh Island, not far from Thunderbolt. After a day on the water, he usually dropped his boat off at his lot, unloaded oysters into the cooler, and went home. Aquaculture equipment sat in piles around the place, hidden from roadways by moss and brush. Having spent months at a time submerged in brackish water, the gear vented off a piquant aroma.

The stench displeased the lot's neighbors, and they complained. Nothing about the smell was alien to Thunderbolt, but the community had changed. The city's authentic bouquet didn't appeal to the retirees and commuters now calling the community home. So Justin did what he could to make the full-time residents happy by washing gear in fresh water before depositing it at the lot. Neighbors, though, weren't the problem this day.

January weather that year was mostly dry. Temperatures spiked well into the seventies through the middle of the month. A wild storm had passed through over the weekend, and Justin worried his cooler had lost power. In a few days, Justin was due in Atlanta for Coastal Day with the oysters contained in the unit. They were alive, technically, surviving because their two shells clasped together tightly enough to create vacuum-sealed sanctuary in which the animal could live out of water for days. But if their surroundings became too warm, they could become unsafe for human consumption. So he popped by not just out of concern for live shellfish, but for the health of consumers.

His truck rolled onto the lot, tires following worn tread depressions in the sandy soil. He walked toward the cooler and was grabbing the padlock when he noticed the thermometer: forty-seven degrees, forty-nine degrees, fifty-one degrees. Today he can't remember the exact temperature, only that the reading was bad news. He tapped the glass-covered dial. Maybe it was a fluke.

Behind insulating layers of galvanized steel and polystyrene, his oysters might be dying. Plenty of them in there, cleaned and bagged. He had grown some of them himself through great effort and dedication. He felt proud enough to show them off in Atlanta. But if this thermometer was correct, the oysters were now trash.

The storm had taken out power to the cooler, and they had been sitting in a warming environment for over twenty-four hours, at best.

State and federal regulations—and Justin, too—prefer that live shellfish be stored at under forty-five degrees, keeping parasites and bacteria on the defensive, and above thirty-five degrees, below which oysters can perish. Justin kept his cooler at forty-two degrees. If the temperature inside the cooler had pushed the upper limit for more than two hours, food safety rules dictated that the refrigerated container's molluscan contents were unfit for raw consumption.

The concern was that *Vibrio vulnificus*, a warm-water bacterium active in temperatures above forty-five degrees, could afflict eaters of raw shellfish with nausea, diarrhea, or worse. Granted, *Vibrio* is more likely to strike via oysters harvested between April and October. But rules are rules, and abiding by them is easier than defending a wrongful-death lawsuit.

He couldn't lie. The result would be disastrous: the lieutenant governor lurching toward the nearest trash can; the Senate whip rushing off for the men's room; an aide dizzy, turning blue.

Justin tapped the thermometer again, then unlatched the door. The air felt off, not quite cold enough. He accepted his fate: he had lost a few bushels of oysters and an opportunity.

He gathered himself and called Joe Maley and Danny Eller, Liberty County watermen who had planned to attend Coastal Day with Justin. Justin told Joe how he had screwed it all up.

Joe laughed at Justin's predicament. Although he used words like *passionate, knowledgeable,* and *energetic* to describe Justin, Joe had seen Justin's confidence undo him before. He meant no ill will toward Justin, but he had to poke fun at the bad luck that befell him. He had to; he was a big, tough waterman.

Joe first met the Spatking when Justin was a grad student. He had come down to Liberty to set up an experiment on Danny's lease. It was a sunny day, warm, but Joe knew that once they were out on the water, winds would cool them down. He warned Justin.

"You might want to get a jacket," Joe remembered telling him.

"I'm from Michigan, I don't get cold," Justin responded.

"He just about froze out there," Joe said.

Despite his laughter, Joe quickly offered to help.

Let me call Danny Eller, Joe told Justin. We'll get you some oysters.

"Of course we were going to help [him]," Joe said. "We weren't going to kick him when he's down."

With a day or two to spare, Joe and Danny cruised out to their leases, bagged up a few bushels, and cleaned them up for shucking in the big city. The crew, which included Dan DeGuire, drove to Atlanta and put on a show in the Capitol. Their audience knew nothing of the drama.

They set up a display in an unadorned conference room: unfolded a table, draped a tablecloth over it, and arranged oysters prettily over a bed of crushed ice. They shared the room with some clam farmers who had made chowder. Dressed in button-ups and ties and even a bowtie (Dan), the oystermen, joined by Dom Guadagnoli, lined up behind the table, smiling and shucking for a growing line of legislators, businessmen, aides, and interns, who filed into the room eager for free seafood. They couldn't shuck fast enough, Justin remembered. Over and again, they opened oysters, offered them to hungry partisans, and collected the spent shells.

An older lawmaker shuffled through the queue. Joe recalled that he looked as if he had spent his whole life under the Capitol's gold-leaf dome—aloof, out of it, nearing senility.

"Are these from Apalachicola?" the politician asked Joe, referring to the South's famous oyster grounds in Florida. Justin described it as one of those record-skipping moments like in the movies. Joe seethed, but everyone just laughed and answered politely. No, these are from Liberty County, sir.

Watermen understood that politicians knew little about their lives. Agriculture under the dome meant commodities: cotton, onions, peaches, peanuts. Ask a political about fishing, and he would assume you meant angling. These guys ate expense-account dinners all the time, Justin and Joe figured, and they probably ate shellfish when they did, never questioning what was on offer from their own state.

"Georgia has a coast?" Watermen and coastal conservationists heard that inquiry more times than was fair. People knew very little about where or how oysters lived, and politicians could be counted among that crowd. That the state's most important citizens possessed a working knowledge of its shellfisheries was too large an ask. But Justin and Joe and the rest were smart to try.

"Is your work being messed up by the water wars?" another legislator asked the group. No, the oysterman made clear, the legal battle over the Chattahoochee River's freshwater supply to Apalachicola Bay in the Florida Panhandle's stretch of the Gulf of Mexico would not affect a few acres of salt marsh 350 miles away that drained into the Atlantic Ocean. The water streaming out of Lake Lanier, north of Atlanta, and eventually into Apalachicola Bay had become the stuff of governor-led lawsuits. Sprawling Atlanta gobbled up all the outflow, slowing rivers to the south to a dribble. For Apalachicola oysters, that meant less freshwater to mix with the salty Gulf.

Oysters grow best in brackish waters—a mix of salty and fresh—and in Apalachicola, where oysters had been the dominant industry for decades, these water wars' main victims were bivalves and the people who made a living from them: harvesters, shuckers, canners, restaurants. It was a fascinating story for anyone concerned about fishermen, rivers, or the impact of urban development on the natural world, but it had nothing to do with Georgia oysters.

With local reporters, Justin showed a deft hand at publicizing his water work. The experience in Atlanta proved that convincing politicians—unlikely but necessary allies in the growth of the industry—required a nuance that he had yet to master. First he would have to raise awareness in the big city that watermen were alive and kicking down on the coast, that clearly delineated yet somehow mysterious boundary on state maps. Next would be to counter both misinformation about farming the ocean and outright confusion, like the route that some of the South's major waterways followed. As representatives of a team, Justin and Joe weren't prepped for the attention. The smiles came easy, but the lethargy of legislators' geographic and piscatory knowledge wore them down. The oystermen would have to build up a resistance to it.

They were still learning how to step up to the podium. Once there, the dense questions wouldn't stop. The meaningless handshakes would only increase. The bureaucracy they would be forced to navigate would meander in ways more convoluted than any estuarine creek. It was easier to sell the marsh in the marsh, the growers knew that, but the coffers were stored miles away from open water. They would have to get used to making the drive.

SUNBURY

Justin Manley on a Skidaway Island dock.

The first image of Justin Manley that I saw came in the form of a digital photograph attached to an emailed press release from the University of Georgia's Marine Extension Service. The news touted the revolution Justin had been fomenting, that a shellfishing boom was soon expected.

In the photograph, sent to me in the spring of 2013, he stood on a creek bank, the tide long gone, a fragile mix of mud and shell beneath the soles of his tennis shoes. He stood as rigid as the phalanx of pines rising at the back of the frame. Legs spread, his head bowed in concentration, eyes fixed on an oyster held by a gloved palm. Powerful, like a hunter at peak skill.

A lone man in the marsh. I compared him to the hammocks of myrtles and saw palmettos, the solitary patches of trees that interrupt vast and flat parcels of grasses. Tidal pools and streams divide him from leathery branches of dryland timber. He belonged, but was still remote.

I called the phone number included in the release, and invited myself down to visit the Spatking shellfishing operation. Sure, Justin said. It was May, and harvesting season had ended, but there was still plenty he could show me.

In the week before our meeting, we spoke regularly on the phone. Justin watched the weather radar like a fiend, and the weekend forecast appeared ominous. When taking novices on the water, he wouldn't risk poor conditions. Worried that lightning or wind would maroon us, he wouldn't green-light the trip until he could promise clear skies.

"The water, it's been so choppy, I don't know if we can get out there," Justin said. "I'll keep watching the radar and get back to you."

He called again with no news. And again the following day, only the prognosis was delivered with a minute increase in positivity. The same worry persisted: "I'd hate for you to come all the way down here and us not be able to get out on the water," he said.

Justin expressed caution from experience learned in real time, and he had heard plenty of treacherous-weather stories secondhand. His oystering peers practiced the skill of knowing when to take the day off. Not that they always listened to instinct.

"You learn when not to go," Joe Maley once told me. "You can take the weather report and wad it up. Our local weatherman is wrong so many times, he should lose his meteorological license. Used to, before I would go out, I'd look at the treetops out there. If they're bent over, I didn't go. Depending on which way the wind was blowing, too. If you got a westerly wind, at fifteen to twenty miles an hour, you know when you're coming back on the incoming tide, you're going to be on some . . . it's going to be rough. The wind blowing at the wave, it could stand you straight up."

The act of harvesting oysters, while hard work, might appear to outsiders as idyllic, with shifts marked by sun and a fresh breeze. But a life spent on the water did not preclude an oysterman from having a bad day. The tragedy of Martin Luther Stewart Sr., a South Carolina oysterman lost in the line of duty a few months before I met Justin, served as one of many warnings for the Spatking.

Born in Hilton Head, a South Carolina sea island about eighty miles north of Justin's oyster lease, Stewart learned to fish, crab, dig clams, and break down oysters along Broad River inlets. For the men who taught Stewart the water trade, these rivers offered careers of bringing catches to shucking and canning houses located at Port Royal and Daufuskie Island. For Stewart's generation, business was okay because of little competition, but you had to diversify income streams. But for men like Stewart, or any of the watermen I have come to know, an income's only purpose was to pay for the next tank of gas, the next hull patch, another fuel filter, or a new set of oars, if times are particularly hard and the motor is busted.

Stewart left home at a young age to make a start as a seaman. He called Key West his first port, then jumped on a shrimp trawler working the Gulf of Mexico. He understood the shrimp trade inside and out, taught by his father and older brothers, so that, as his obituary stated, "he too would be able to making a living and provide for himself and his family." He followed the fleet along southeastern coasts, making his forebears' fishing craft his own. Eventually, he came home to the waters he knew best: the Broad River, Port Royal Sound, and a hundred more creeks nameless on most maps.

"He could throw a cast net right out his back door and bring in baskets of a variety of seafood," read Stewart's obituary. "He loved the life of an outdoorsman and was an adventurer at heart. Martin loved the sea and all of its bounty."

But the sea had a nasty side.

On the day after Christmas, Stewart and a fellow oysterman named Vincent Chaplin loaded down a skiff with twenty-five bushels of oysters. They left a deepwater dock on the Broad River at 1:30 p.m.

Christmas week had been the wettest week on record in South Carolina since August of that year. And on Boxing Day, it took another unpleasant turn. Fresh gusts of wind whipped up that Wednesday, measuring at airports around the state at forty-four miles an hour.

The water that Stewart and Chaplin worked was definitely choppy. Gale-force winds don't make for easy seas. Forty degrees, which felt even colder with the winter chill that blew in that morning, would have been plenty frigid and uncomfortable for these Lowcountry natives.

They had spent the afternoon working on the other side of the river from the dock, harvesting along the banks of Corn Island. At about four, the two watermen decided to cross back over the Broad, a journey of less than two miles.

Stewart's skiff surely took broadside waves that rose six to nine feet. Add the wind, and it was enough to capsize the vessel. Sometime between four and five that afternoon, they flipped, men and oysters tumbling into the river. The bushels sank to the bottom, the men floated in their life jackets.

Chaplin and Stewart attempted to swim to shore, but wind and waves forced them back to the overturned boat. Chaplin clung to the skiff's metal sides. In all the effort to stay afloat, Stewart lost consciousness and drifted away on a wave.

Over the next three hours, the skiff floated across the Broad, nearing the boat dock they had set out from that afternoon. Rescuers were called out when Chaplin and Stewart failed to return by five, as promised. Two hours later, they found Chaplin alive and holding on to the hull.

They found Stewart, too, just seventy-five feet away from his boat, dead from hypothermia and lungs full of water.

Stewart's death shook the fishing community, and news of the tragedy caught the attention of oystermen one state to the south. Such calamities weren't common to watermen, though many tell of near-death experiences. Taking such risks was part of the job. Even though one of their own had paid dearly for working the water in harsh conditions, Stewart's death didn't cause any of his colleagues to rethink their career choice.

From his hospital bed the following morning, Chaplin called the owner of the seafood distributor for whom he and Stewart fished.

Won't be in today, Chaplin told his boss. But he'd for sure be in tomorrow.

Reports had improved enough by the weekend that Justin approved of my making the drive to Savannah. We rendezvoused in a Starbucks parking lot, where he waited outside a white v-6 Ford F-150. He seemed to not notice that he had stepped his permeable sporty footwear into a puddle. He wore a T-shirt, and crammed his hands into jean pockets as his only brace against wind. I had prepared myself with a wool sweater and a Gorton Fisherman's rain slicker, yellow as table mustard.

"I hope you don't mind getting wet," Justin greeted me.

Rain had pelted the coast through the night. Low gray clouds made our boat trip seem improbable. But the rain would dull, Justin promised. He assured me that by the time we pushed off into the water, the last mutterings of the storm would quiet.

Justin didn't live far from the parking lot. His family owned a condo nearby. From here, it took nearly an hour to reach the dock closest to Justin's oysters. Most watermen live quite close to their oyster leases, certainly in the same county. But regulators knew Justin would be a dependable steward of his allotted marsh, given the depth of his shellfish knowledge and experience, so the Department of Natural Resources made an exception for him. The favor wasn't a total win. Miles collected on his odometer and gallons of gas rung up on his credit card took their toll on his bottom line.

We drove out of town, headed south on I-95 toward Liberty County, and took the exit for Sunbury. Historians call Sunbury a sleeping city. Even worse, a dead one. They say this because Sunbury before the Civil War had bustled as a port, exporting resources from interior forests and farmlands to Europe. None of that remained. The community napped. People here desired a relaxed home, the coastal life. They sought seclusion, content to let folks in Richmond Hill to the north and Darien to the south flirt with urbanity.

Retirees made up most of Sunbury's year-round population. Others found work in Chatham and Bryan Counties and slunk home to hide and sleep. Whatever was original here, built by European hands beginning in 1758, had crumbled long ago. Tufts of wiregrass hid evidence of the foundations of colonial homes. Safe behind a wrought-iron fence stood the gravestones of early settlers.

What was new looked less than fresh. Condo façades had composted quickly into the landscape, given a gray and green patina from splatters of sea air. BMWs sat idle in the driveways of low-slung residences. Despite the recent construction, dirt paths connected some houses to a main road. Where asphalt was laid flat, sand-pecked earth crept in and cracked the streets.

Nearby Fort Morris, a Revolutionary War–era battlement once called Fort Defiance and later filled with hundreds of soldiers in the War of 1812, still attracted tourists a few days a week.

A marina hosted shrimp boats, two commercial vessels, and pleasure craft large and small. It was operated by the Sunbury Crab Company, a dockside restaurant serving fresh-caught crabs steamed and delivered table-side, claws intact.

People called Sunbury home, though not too many. But for a time, Sunbury had rivaled Savannah in population and industry.

The botanist William Bartram visited Sunbury in 1774, about sixteen years after Congregationalists from Dorchester, South Carolina, moved south to establish a new settlement, with blessings from the English Crown. Bartram found little to record. He took up lodging in a house built among the town's four hundred lots, set on a street grid just like Savannah's. He rented a horse and explored the pine forests set back from oak-dominated river bluffs. He chartered a ship and observed the nearby sea islands, Ossabaw and St. Catherines. In his report to King George III, he discovered "nothing new, or much worth your notice." There were oysters, he told the king, a fair amount, but in no greater quantity than in the Carolinas.

From colonial days to the Civil War, Sunbury bustled as Georgia's second-busiest port, the site chosen for a long bluff that formed a natural dock at high tide.

Button Gwinnett, businessman, politician, and a representative of Georgia at the first Continental Congress, frequented Sunbury, boating in from his home on St. Catherines Island, where he was trying out a new profession, planting. His crops—cattle and cotton—fared poorly.

In the late 1700s, fifty-plus commercial ships a year navigated the Sunbury Channel, each ship landing a small victory in the city's fight to attract naval business away from Savannah. The ships left port with rice, staves and shingles for building, lumber, cotton, and cattle.

In the 1830s, the last commercial packet ship left Sunbury, bound for Europe. Sunbury's last dockmaster, James Holmes, recorded its departure in his log. He watched the cotton-laden ship skirt the island that splits the Medway River into two channels, a steam engine aiding its departure. It then gained speed through St. Catherine's Sound, past Ossabaw Island, bound for Sweden. Within two hours, long after Holmes had quit paying attention, the ship disappeared on the horizon.

Parts of an old cross-state highway that connected Sunbury to inland cities can still be navigated. Along the graveled Old Sunbury Road are hunting lodges, game preserves, and, as the road nears Interstate 95, an industrial park in which a factory outpost of a global aluminum producer shares utility connections with a shipping hub of an international retailer.

Apart from Fort Morris road signs—some wooden and painted brown with white lettering, other bronze and aging green—no sign exists of Sunbury's past life. Sunbury disappeared long before General P. G. T. Beauregard blasted the Yankee troops holed up in Fort Sumter, kicking off Civil War fireworks.

Today, tall oaks grow within the brick footprint of a nineteenth-century health spa. Hints of the red foundations, though, can be seen only from the middle of the Medway River.

Given land by General William Sherman after he took Savannah, freed slaves formed an upland community called Seabrook. Houses built by their descendants line the road into Sunbury, stationed near two Baptist churches. The canopied Old Seabrook School Road joins mobile homes to the main Sunbury road. At that intersection sits a living history museum devoted to antebellum slave life.

Justin named his choice oysters Sunbury Selects, but the deep history of the city did not influence the decision. It was a matter of convenience. The Sunbury public dock was the easiest place to start the boat journey to his lease. He didn't need Sunbury for its name—any local name would have worked. Seabrook. Jones Hammock. Medway River. Justin needed a place from which he could build an origin story. It was about marketing.

Oyster-growing media in the mud near the Crooked River.

Justin checked his watch. To get a good look at his operation, we would have to boat out on the outgoing tide, so we still had some time to kill. We drove to a secluded homesite owned by the waterman Danny Eller, who was not around at the time, where Justin parked the *Green Hornet*. Justin, Danny, and Joe Maley worked Liberty County's three oyster leases, located around the perimeter of St. Catherine's Sound. Danny's house sometimes served as an outpost for the Spatking, both locker and staging ground for workdays. From Danny's, we went to Joe's house. If I wanted to get a sense of what was happening in the Georgia oyster industry, Justin advised, I needed to meet watermen, not just see oysters up close. Joe was a good start.

Joe lived just off the road into Sunbury on a short, sleepy drive that skirted Fort Morris State Park, bent hard right around Joe's homestead, and dead-ended near the edge of a marsh bluff.

Joe's baby blue home took up a fraction of the two-acre lot he had owned for about a decade. Young trees and shrubs governed much of the footprint. To build the house, mature pines and oaks had been uprooted, and new plant life had been slow to fight its way back along the ground. It was a house afloat in a pool of Bermuda grass, although the tops of the few trees left rooted did their best to block the sun from warming the tin roof.

Two decommissioned boats sat in the cleared pathway. One, a yellow skiff, had served as Joe's first shellfishing vessel. He had learned that western winds flipped the boat with ease, so he retired it as soon as he had the money for another.

The other lawn-ornamenting boat, a red-and-green-hulled cuddy cabin, had disappointed him equally. He had bought the fourteen-footer because of its fiberglass pilothouse and deep hull. The former would protect him from winter winds, the latter from aggressive waves. Unfortunately, caveat emptor, the hull came cracked and proved unfixable. It became the most expensive yard art Joe had ever purchased.

Busted or inactive, the boats still served a purpose. They offered Joe the opportunity to lob self-deprecating jokes about his inability to navigate light vessels in queasy seas or spot a lemon when it squirted him in the eye.

Joe, sixty-three, was the current president of the Georgia Shellfish Growers Association and had been one of Justin's trusted advisers. Silent, industrious, and sarcastic, Joe embodied the watermen characteristics that Justin admired.

Joe didn't start shellfishing until the early 1990s. Time on the marsh had been a hobby begun in the U.S. Army, when he worked as a construction inspector at Fort Stewart, nearby in the Liberty County seat of Hinesville. On weekends, Joe retreated to the marsh to collect oysters and clams for fun and a little extra money. As retirement neared, pastimes increased in importance. Now, he bothered only with boats, shellfish, and other marine projects.

Justin rapped on the front door. The knock woke Joe from a midmorning nap, so he answered with half-lidded eyes. Joe was a boulder, a stout and short man for whom deliberate movement progressed at a centenary speed. Buzzed gray hair ringed his head. I guessed he had fallen asleep in the recliner positioned near the door, lulled into dreams by a TV news sedative, the program still humming as we entered. Or maybe a tranquilizing paperback had knocked him out, one from the bent-spine collection piled on an accent table.

Opening the door wide for us, he scratched his belly with fingers thick as bratwursts. Their girth matched Joe's portly waist, but I knew his digits' density to be related to labor and not diet. I had shaken enough hands with working people to know that a life spent turning wrenches or pounding nails builds muscles in the unlikeliest of places, fingers chief among them.

This movement, regular scratching at his lower stomach, would become a tic that endeared Joe to me. When he wanted to exude mellowness, in his living room or on the marsh, he rubbed at his belly. He was ready to chill, he seemed to say, or had just finished napping. Either way, there was no hurry. We were just waiting on the tides.

Did we want coffee? Yes.

"The future of the industry isn't economically secure," Justin said. "This spat stick, it's just a kick start, a poor man's method. It's not meant to be the industry. It's ready to bust open, but it still needs some sweat."

He sat away from the table, giving his legs room to stretch, and bent toward me, elbows perched on his thighs. Justin seemed to be providing the sweat, but what could make it secure?

Without a beat: "A hatchery."

Oysters grown in a lab—a hatchery—until they could safely survive in open water would create a reliable supply of bivalve farms, he said. Hatcheries provided predictable quantities of seed at a predictable size and price. Compared to wild seed collection, which relied on nature's dice roll, hatcheries were more dependable. Georgia had a successful clam industry, he explained, fed by a hatchery in Florida, and watermen had built dependable incomes from clams. Farming clams was far easier than oysters, he said, but the product garnered less than half the price a farmed oyster could. With just a few simple variations, he promised, oysters would be as successful as clams.

A regional hatchery, Justin added, would provide the "foundational economic stability" to help the industry grow. A hatchery required a growing stable of clients for its own success, and farmers required seed on demand. Loyalty was mutually beneficial.

The more familiar I became with the oyster industry in Georgia, the more I realized just how right—and how wrong—Justin was. Those variations he discussed seemed simple to him, but change came hard and slow with these fishermen, even those who had joined the association.

One thing I would come to understand about oystermen in Georgia is how much they enjoy the simplicity of a wild harvest. Little investment, not much long-term planning. Check the weather, check the tides, and head to work. I didn't meet watermen desperate to deplete a natural resource. They labored slow and with purpose. Still, there was a freedom to what they did, a choice to go or not to go, work or not work. A farm, they certainly knew, tied them to routine. Such flexibility is not so easily discarded.

One thing I didn't understand was the necessity of the hatchery. There are oyster hatcheries in the Northeast, on the West Coast, even in Virginia, I said.

"Can't you order spat from one of those?"

"Sure, but it wouldn't be a Georgia oyster. It wouldn't be from this place. And that's important."

Regulatory concerns also slowed oyster seed from crossing borders, I learned. Introduced spat might carry disease, or a northern brood line might not be genetically suitable to the South. It wasn't comparing apples to apples, but Braeburns to Arkansas Blacks. In the genes, vast quantifiable differences existed.

To simplify it all, Justin pitched a passionate quote: "You eat oysters to connect with the ocean, and most of all we sell place, one little corner of that ocean."

Most wild oysters on the eastern coast of the United States are the same animal, *Crassostrea virginica*. But place defines taste. Justin wanted an oyster bred to retain the animal's natural resiliency. Wild DNA made replicable.

I wondered how likely, or how quickly, a hatchery could materialize. The University of Georgia could do it, just like land and sea grant institutions in Louisiana and Alabama, Justin said. Or a private business could come in, but a company wasn't likely to invest in such infrastructure until it knew there was a viable market. What was the rush? I wondered.

"We need to develop the industry before development picks up," Justin said. It's a race, he explained, and many competitors had better odds of success than skiff-driving oystermen. Suburbanization crept out every year from Savannah and Brunswick, threatening to squeeze rural towns like Sunbury. Increasing interest in offshore drilling had begun to worry coastal conservation groups. Near the houses of two oystermen I would meet in Camden County, plans to build a spaceport that would send rockets far above the marsh wash progressed at the behest of local government. A deep-pocketed gas and oil company called Kinder Morgan had proposed the Palmetto Pipeline, a coast-hugging spur of the company's Plantation Pipeline that would move refined petroleum south toward Jacksonville, Florida. Nobody liked the idea, even the state government. The state and environmental groups waged quite a fight against Kinder Morgan to save the marsh, wetlands, and rivers from an oil spill, but now that oil interests had come calling, it was certain to not be the last battle.

Let the marsh build its economic worth sustainably, Justin argued, and maybe then fishermen could leverage their commercial impact to prevent threats to their livelihoods. If it came to that, the industry would need leaders. Justin had already positioned himself to be one voice. He hoped the GSGA would also speak, collectively.

Through all this conversation, Joe sat mostly silent, rising from the table to pour coffee from the blown-glass Chemex into mugs.

I tried to wrest words from his sleepy mind. What do you think the industry needs?

Workers, Joe said. He was only getting older, and oystering chores weren't getting any easier. He rarely found anyone younger than him to stay out all day picking oysters and then come home wet and wind whipped.

"Casual labor is hard to come by," Joe said. "The sons of bitches got to show up at 6:00 a.m."

Joe swore under a veil of toughness. He winked as the curse left his mouth as if to say, I'm just putting you on. Or maybe he looked past me, behind the couch to a wide window ledge where images of his grandchildren glinted behind glass frames. A toy train carrying the word *Granddaddy*, each capitalized letter on a separate car, weaved through the photographs. The third wooden *D* had tipped over. Perhaps he wanted to tell the kids sorry for letting such an expletive fly.

Too often, Joe recalled, he hired on help that worked one day at the most, earned a day's pay, and never came back. Wimps and assholes, he thought, but he'd never say that to their face. A labor shortage wasn't just Joe's problem. Rather, it plagued all fishing business along the coast.

Factories and urban centers lured away most capable bodies. Not everyone was like Joe, desirous of a spartina-fringed corner office. Rap sheets, more often than not, trailed those drawn to the off-grid cash pay of water work, and dependability was not a trait picked up in county jail. Here was the problem as Joe saw it: if he couldn't get workers to pick wild oysters, forming a team he would trust enough to tend an oyster farm sounded impossible.

I looked at Justin.

"It's a problem," he said. He could offer no foreseeable solution to the labor troubles, so he shrugged. "But I believe once people see the money that can be made farming oysters, it'll get their attention."

Justin checked his watch again. With a nod, he motioned that we needed to leave. We said our good-byes to Joe. It was time to catch the tide.

THE SPAT STICK

An oar on Daniel DeGuire's boat.

A lone fisherman standing on the L-shaped walkway that stretched over the Medway observed as Justin and I turned into the Sunbury dock parking lot. His rod leaned against a handrail. His gray sweatpants sagged, and his round belly filled a black mixed-martial-arts T-shirt promoting a popular brand of fight training gear. He returned his gaze to a smartphone screen, his wispy white line blown in an arc by the breeze. He gave us little thought, glancing at us and returning to his digital meditation.

Justin backed the boat trailer down the ramp into hickory-colored water. Trailer chains clanged and the frame ached. The hitch juddered as the trailer bumped over traction divots cut into the concrete.

Justin jumped in the boat, and I unhooked it from the trailer. The boat slid into the Medway as Justin ignited the *Green Hornet*'s motor. He puttered in reverse and roped the boat to a small dock accessible by a staircase off the walkway where the lone fisherman worked.

I parked Justin's truck nearby and walked the gangway to meet him. The aluminum lane pinged lightly under my footsteps, like a child tapping a penny against its frame.

Underneath me, below the aluminum path, slept ballast stones from Sunbury's port city past. Before steam engines, ships bore these heavy loads while crossing the Atlantic to weigh down their empty holds. Only yards away from the public dock was the plateau where Sunbury's colonial inhabitants built their shipping outpost. Lasting evidence of the town's former might sits underneath low tide.

Justin had dressed up since our meeting in the parking lot. A fleece half-zip sweater covered his torso, and a heavy beige skullcap was pulled over his shaved head. Girded by cotton and wool layers I had left my fisherman's jacket in the truck. The rain had begged off, but heavy clouds hadn't. Soon we were zooming up the Medway and into St. Catherine's

Sound, saline gusts rouging our ears and noses. Justin used two fingers to control a box-mounted steering wheel, held up by the gunnel and a metal arm on the starboard side. He stuffed the other set of five in a pocket for warmth. He stood over the wheel, leaning in to reach down to it.

A cold wind bit through his stubble. He smiled. Sporty sunglasses, temples tucked under the winter hat, hid his eyes.

We buzzed through a solemn marsh. The gray clouds diminished the glint of spartina, the reflection of river.

Justin's control of the skiff impressed me. He knew what waited behind each bend in a channel. He had crammed charts between his ears, could recite coordinates on the spot. He divined direction from the churn of the tide. Nautical charts had been committed to memory. To work a commercial lease with any efficiency, he had to. A workday spared no time for course plotting. The commute from dock to lease was short and had to be made with a semblance of speed.

In the few moments on the Medway that morning, his command of the boat was obvious: I motioned that I had a question to ask, and he slowed the motor without jarring me.

I yelled over a sputtering motor: "How do you know where you are, where your lease is, where the river ends?"

He reached for his wallet. From a sleeve he slipped out a laminated card, about two inches tall and four inches wide. A Master Collector card, obtainable only when a lease was granted by the Georgia Department of Natural Resources. With one hand steering, gripping the wheel now, he held up the card. I leaned toward it without standing up from my seat on the foredeck and saw a pixelated mug shot of Justin, his face organized beside vital details. It was not unlike a driver's license but simpler, fuzzier, less designed. Purely functional.

Every leaseholder who worked the coast had a card just like Justin's. It made him an oysterman in legal standing with the Department of Natural Resources, and it meant the Department of Agriculture approved of him as a legal seafood supplier. It meant he understood how to handle shellfish in a manner that ensured public safety. It documented that he had established an onshore site to process and store his catch. The department's demands weren't ludicrous, but to meet them a waterman had to invest in a commercial cooler and build a cleaning facility. Other than a boat, those were their biggest expenses.

Once certified by the Department of Agriculture, shellfishermen went to the Department of Natural Resources. This division's chief condition was that every bushel of oysters harvested and sold commercially had to be affixed with a shellstock tag. On this tiny document are the dealer's name (or harvester's name), his certification number, the harvest date and location, what type of shellfish was caught, and how much of it. At the start of oystering season, in October, the DNR takes a water temperature reading. If it is above seventy degrees, *Vibrio* bacteria are plentiful. But it's just to be safe. By the fall, marsh waters have cooled enough to disperse congregations of such spoilers, but they still have to fill out a tag.

The Master Collector card meant that a committee, not unlike the board of a commercial bank, had green-lighted Justin to work a corner of the state's beloved marsh. If the committee approved of Justin, it meant that the head of shellfish operations at the DNR's Coastal Resources Division, Dom Guadagnoli, approved of his character and his financial stability, and didn't consider him a thief.

Dom's division also permitted a secondary card, called a picker's permit, for employees of master collectors. It allowed the holder to pick only from a specific lease. A picker must have a fishing license and must not have been arrested more than three times. He or she also had to watch a video about sanitation at Dom's office.

Run this gauntlet, then it's off to the water.

Justin slowed the skiff so that it no longer flopped violently against waves. We jounced awkwardly now, but I could focus on the tiny piece of plastic in his hand. He flipped the card to the other side, revealing another purpose of the ID: a map of his shellfish lease, a small abstract representation of his personal marsh.

Dark green forms illustrated the mainland. Gray blobs, filled in with vertical hash marks and tiny crosses, marked the grass-topped marshes through which tidal creeks funnel ocean water. Horizontal dashes, a shade of gray in a mess of blue, indicated the wet stuff. Two separate red lines— shaped like two odd hexagons—marked off the two zones that made up his lease. Lean creeks and a wider river washed through all of it. We would soon be rushing up the former. But first I wanted to get my bearings on this map before Justin blasted the motor again.

The red lines on Justin's collector card were the opposite of arbitrary.

Many factors determined the health and survival of oysters. One is entirely anthropogenic: pollution, mostly in the form of human waste. The monitoring of fecal coliform performed by Dom and the DNR determined the placement of a lease. Dom's agents also searched the marsh daily looking for signs of wildlife that had expired there.

Every three years, Dom and his crew staged a sanitary survey to assess new or emerging pollution sources, like new subdivisions. Every ten years, they ran a full-blown assessment. No creek was left untested. Beyond that, the feds had their own tests.

Only a few areas have been completely shut down for shellfish harvesting. Savannah, Brunswick, and St. Simons: three of the most populated locations on the coast. Too many people there and not enough water flow to dilute human inputs.

The State of Georgia owns all its riverbeds up to the high-water mark. Unless pollution makes it untenable, and as long as there is shellfish nearby, the DNR can lease out any stretch of marsh it wishes. The only outliers here are Crown grants—legal ownership of land and water given by the king of England. In the twenty-first century, authorizing a Crown grant requires an accurate reading of a handwritten document composed three hundred years ago. And the attorney general has to agree with that reading. Dom could count the number of valid Crown grants on one hand, and only two of those are involved in shellfish. None of the others were expected to start harvesting oysters.

In some North American estuaries where oyster farming is prevalent, coastal homeowners can see oystermen at work. Not in this marsh, though, and that won't change soon. Leases like Justin's are remote for a reason: safety. The more isolated the lease, the cleaner the water and the more bacteria-free the shellfish. Remote oyster beds, hidden from most eyes, help the DNR ensure public safety.

To watermen, red lines also inscribed primal boundaries in thick mud. They were borders of stewardship, one man's peninsula of duty, borders of ownership that should be crossed with respect. Pass through the lease, that's fine. Cast a line if you must. But don't touch the oysters.

Sectarian angst doesn't come naturally to Justin. He had heard enough stories from watermen to understand that he was not as territorial as they were, but he'd spent enough days on the water to grow a bit leery of strangers.

We bobbed in the middle of the Medway, and I didn't dare pick the card from his hand. I worried what the consequences would be if an unexpected hull wiggle unfastened the card from my fingers.

"Most guys would kill you before they let you see this," Justin said. It was half a joke. Prowlers who think nothing of taking oysters from places they shouldn't would make good use of the map. When an oyster bed was plundered, half the time the victim knew the thief. Two security measures that an oysterman could take were silence and seclusion: the less that others knew of your shellfish kingdom, the better. That poachers navigated creeks just as well as oystermen poised another problem.

One McIntosh County waterman told me that disputes over shellfish were an extension of feuds from dry land. "It's just a small town," he said. "We all hate each other." There was a line between a fisherman who works the water and a no-good redneck who steals from others, I was told. What the law says about which oysters belong to whom isn't necessarily enforceable or even recognized, another waterman told me. It's about people and their relationships. Arguments, to cut right to it.

"The thing about the coast is that the histories run deep," a waterman told me once. Favors, and slights, are not soon forgotten. An upland dispute might see retaliation on the water in the form of a theft.

Limiting public knowledge of oystering areas was essential for the farming methods Justin had set up on his lease: the sticks, the racks, the bags all hung buffet-like for oyster poachers. Deeper now in the marsh, I saw the operation's vulnerability.

The tide pulled away from the high-water line to reveal spartina and blooms of black mussels growing where white stalk met muck. Clumps of oyster shells began as the bank sloped toward the river bottom. The deep tides offered oysters hours out of the water. The respite thickened their shells. As the tide washed out of the marsh, oyster pursed their lips, held their breath, and halted their water filtering and algae eating. Parts of the marsh were flatter, like wide platters stocked fully with bivalves, a carpet on which waterman boots can comfortably crunch. These were choice workspaces to crack and cull through clusters.

Along the marsh banks, I saw gray mats of oysters laid out like sheets of plywood. I couldn't have walked here, even at low tide. The all-sucking mud would have eaten my shoes. Long and fallow pauses ran between slabs of oysters. Only a boat could connect a fisherman between bunches.

Unnatural objects appeared. Thick wooden posts spaced evenly apart stuck out of the water near the banks. Bundles of thinner sticks poked out of the bog, their appearance less explainable than the posts, which might have, I assumed, marked crab traps.

All were stages of Justin's aquaculture project. The fat and thin stalks were a section of his shellfish grow-out area. What I couldn't see from the boat were the lines of rebar trellised between the newels.

On those bunched shafts, Justin collected oysters in their young, microscopic form. He called them spat sticks.

There was no difference between the pipes used by plumbers and Justin's device. Some of the bundles Justin had set out were less than one inch in diameter, others thicker. Some white, some gray. A season of heavy oyster set had ashened the white staffs. Gray and white bands pocked the sticks like a ringworm invasion; these were crumbly, hard chalk traces of where baby oysters had landed.

Justin had jammed long sticks of PVC into the mud. When oysters began to spawn—in late April or early May, and continuing in spurts through the summer—they sent sperm and eggs, depending on their sex, to float on the marsh stream. Sperm met egg somewhere in the current, and the couple, in the form of a new larva, found a place to set, the stiffer the better. In the wild, a new larva's home is often something hard, like an affixed piece of driftwood, or even the same bed of oysters from which the new offspring's DNA originated. It could set on its own ma and pa. The spat stick caught the progeny.

The system behind the spat stick was simple: Justin seasoned bendable PVC, about a half inch thick, in ocean water for a month or two before spawning season. He drove the stick deep enough into the marsh that it would see some exposure at low tide. This let the spat collect on the stick over a season. Once the new oysters had set, the white plastic turned drab, like barnacled reeds. Covered in oysters, plastic looked native.

After the oysters had grown to a quarter inch in size, maybe a bit smaller, they could safely be removed en masse from the spat stick. During the last set, Justin would collect a bushel of sticks and take them back to the mainland for processing. He laid out a plastic sheet and a tub of salt water to catch what popped off the seasoned plastic. He jammed one end of a stick into the ground and then bent it from the top. The middle would curve out plump like a distended belly. Some of the young oysters would hop right off from the pressure. The stragglers were slipped off with a pocketknife. He then graded out the oysters by size—at this point few were larger than a half inch—separated them into mesh bags, and put them back in the water to finish maturing. They would be graded again later. If oysters survived the next stretch, the struggle was over. They had made it, but then it was up to the farmer to keep them alive. Justin used pearl nets hung between posts as a nursery.

In the bags, oysters were relatively safe from predators. Mud, though, posed another threat. As the mesh bags hung from the rebar trellises, swells of silt would wash over them daily, threatening to choke off the animals' food supply. Justin had to shake the bags at least once a week, which required gas and time—the only two human factors that can be regulated in the aquaculture game. If there was a catch in Justin's oyster farming pitch, this was it. When a wild harvester headed out into the marsh, he expected to return with a few bushels. Burning gas to garden a crop sounded laborious, watermen told him. If Justin expected to sell watermen on farming, he would have to make all the work worth it.

Collecting oyster spat on a surface was nothing novel in the oyster business. Without spat collection, bivalve availability would be limited. Humans have long caught oysters with a variety of materials, from bamboo to stone to ceramic tile.

During the Han dynasty (206 BC–AD 220), Chinese fishermen let oyster larvae set on stones and shells placed within the intertidal zone of their country's long shoreline. Closer to the modern era, these oyster farmers turned to cement rods three to four feet in length. Farmers stuck the cement sticks upright in the water bottom. Oysters would set and then fatten over a period of four years. At high tide, farmers tonged oysters off the sides of the rods. At low tide, farmers removed them by hand. To cross the mud flats when the tide was out, farmers used a "wooden horse," a long flat board that the pilot stood on with one foot, using the other to coast himself along. Somehow, a man's weight skimmed the pliant surface.

To this day, some Asian oyster farmers employ sticks of bamboo as both spat collectors and nurseries. Three-foot-long stakes are buried in the sand for weeks or left in the sun to dry out, and then are soaked in the

Wild oysters settled on PVC structures.

water to attract a sheen of barnacles. The barnacles are removed, leaving a chalk layer that is a perfect substrate for oyster growth.

In the West, two men pioneered oyster farming: Victor Coste in the 1800s; and before him, Sergius Orata, circa 97 BC, in the Roman Empire.

Hired by Napoleon III, Coste used stones and bundles of sticks as spat collectors. He helped France's dwindling oyster supply rebound, ensuring the beloved Belon oyster would survive for generations.

Orata, in what would become common practice around Europe and the New World, cleared the bottom of a lake of debris and transplanted oysters from nearby spawning grounds onto the bare base. Workers shuffled them about, allowing plenty of room for the oyster shells to plump. The action cleaned the oysters of parasitic fouling and silt. To market came a beautiful, buxom oyster, hefty and free of contaminants.

Such delicate grooming of oysters is a daily chore for modern aquaculturists. But Orata was not just an innovative oyster cosmetologist. He could entrap spat like a spiderweb snaring flies.

When Orata's oystering minions spread out another pack of displaced oyster migrants on a fixture of stones, they encircled the mounds with bundles of sticks and tree limbs, catching the progeny of the next spawn.

Orata also developed a system of rope trellises, sometimes hanging from the decks of the villas he rented out. He had oysters affixed to twine cables, likely made out of Mediterranean sea grasses. This hanging culture of oysters was kept in practice in Italy into the twentieth century. Similar methods are in common use among Asian oyster farmers. Mussels, another bivalve that Orata cultivated, are commonly raised in this vertical fashion.

In the centuries after Orata, science codified oyster-farming methods. Bundles of reeds became cement stakes. This concept of catching oysters even trailed down to the southeastern United States. Photos from North Carolina in the 1930s show a circle of paper tubes slathered in cement slurry being stabbed into the low tide by researchers. They are bent as radials toward the middle, forming a round pyramid, a high-tide trap for spat. The practice did not enter common use.

In the early twentieth century, indoor hatcheries began to replace plein air breeding grounds. What once occurred freely in the water was now induced in a climate-controlled vat by laboratory technicians.

But some oyster fisheries were slow to adopt contemporary techniques. They relied on nature to maintain abundance year upon year. If spring spawns were sterile, the fishery would flounder, leaving watermen without income.

In Georgia, there had been calls to bring the oyster industry into the modern era. The last one rang in the 1960s. Not much came of it. Watermen didn't seem to mind relying on nature for continued harvest. Marsh oystermen sold roast bushels—that was it, and that was fine. Their customers hadn't asked for anything else.

Justin arrived in this setting—Georgia without oyster aquaculture—and salivated. A pristine estuary, pure water quality, but an industry stymied by its own mud-plodding habits. University-based hatcheries in Louisiana, North Carolina, and Alabama, all sites of historically large oyster industries, revved up production through the 2000s, and Georgia could be the sleeper hit of a southern oyster revival, Justin thought. He didn't have a fancy hatchery, at least not yet, but he had this "poor man's method," as he called it, that might jump-start it all.

Growing oysters sounded simple: catch some spat and keep them alive. But change shocked Justin's watermen colleagues. In his defense, anything that required spending money shocked them. Investment in gear like the PVC, seven-dollars-apiece bags, and tumblers at twenty dollars each was but one barrier Justin had to convince them was worth breaking. The amount of tending—the farming—required to keep the stock alive, was another. Watermen were used to heading out to an oyster mound, harvesting a wild crop from it, and heading back to shore to shell the bushel. Over a season, the spat sticks and oyster farming required effort. In a word, labor. For the aging waterman population, that effort wasn't worth the money.

In explaining some watermen's objections to oyster farming, Justin laughed. "They think it's too much work," he said. "But what they're doing now is so much more difficult."

He wanted to show me the difference.

Justin steered us north. We left St. Catherine's Sound and shot up a creek nearly wide enough to be called a river. Our path bore us toward the center of one of the red hexagons transcribed on Justin's master collector card. As we curved around a bend, I saw more spat sticks and post sets.

Justin spied a stretch of marsh bank unsheathed by the waning tide. He noted the spot's flint-like hue, a darker shade than the muted sable mud. At a distance, the difference in color hardly registered. We approached the bank and the shading explained itself: the hoary mat was a coat of wild oysters.

Twenty feet from the mound, Justin killed the motor, pulled up the propeller, and grabbed a short paddle. He lay belly down on the bow and leaned his upper body over the edge. He lapped at the river with the paddle until the hull grazed mud. He then dug the oar's tip into the mud to push us closer to the bank. With a crunch the bow stopped on a mess of shell. He tossed a small anchor into the mud, and it landed with a memorable flatulence.

I saw oysters growing skyward like church spires; their ember color insinuated something less holy. Tips of shell shone white, a sign of expansion. I witnessed the famous clumps that earned wild oysters their reputation. A large oyster held the center, pointed toward the sky. Other smaller shells glommed on its back to form the cluster. It flowered with divergent petals. Mud hid in its radiant crevices. There was nothing symmetrical, much less appetizing, about it. I wondered how, if at all, something edible could emerge from this gnarled set of claws.

Justin and I both wore tennis shoes. He warned of the mud vacuum that eats feet and legs, the suck that had hindered his exit that evening when darkness crept up on him. Walk on the oysters, he said. Steeples of them composed much of the bed. But loose ones, worked free by Justin during previous attacks on the bed, lay flat around the center like sunflower fronds. Stepping around the pressed exterior made it possible to walk without being consumed by mud.

Some watermen wear white shrimping boots—plastic footwear that runs halfway up the calf—when out in the marsh. They are called bayou Reeboks in the Gulf of Mexico, but I could see how they would be of no use here. The mud can swallow a calf's worth of a man's leg.

Justin swore by his battered sneakers. He preferred retired running shoes. He never threw away a pair, holding onto them for a cost-saving coda in the marsh. Since there was no extra pair of shoes waiting for me back on dry land, I tiptoed.

From the skiff, Justin lugged a bucket, a knife, and a piece of flat steel about a foot and a half long. He palmed a cluster, the tallest tip likely poking the meat of his thenar muscles (at the base of the thumb), and jostled it free. He held it up for inspection, holding the cluster from its base. It looked like a rigid stalk of seaweed. As many as eight oysters lived on this batch, though most of them were too small to eat.

The clumped form suggested a spine, but none was actually there. Oysters spurred from the middle like prickly lettuce leaves. Newer bracts fanned out from the center, unguarded and assailable. Not all survived a year or two, succumbing to birds and crabs. The harvester of all this mess found stout and salty oysters waiting within, a prize for clearing away the deceased and feeble.

Holding the mass in the palm of a gloved hand, he chopped at it with the steel bar, itself about a half inch thick and weighing at least two pounds. Rubble flew, lopped off in a splatter. (He stepped away so that I wouldn't catch any shrapnel.) The thinning clump bounced in his hand as he flipped it. He tossed it, caught it, and gave it another whack.

The produce of the effort was a bouquet: a shaft crowned by three petals. This is good enough for the sack trade, he said, good enough for oyster roasts.

Justin was adept at this work, even though, as a farmer, culling like this wasn't part of his routine. He stood as he performed this example task for me, but to fill a bushel or two or ten would be an all-day affair. At one point, he bent down at the edge where oyster mound met river and chopped shell into the water. Maybe he could sit down on a bucket or a milk crate. Something.

Justin offered me the steel bar. I grabbed a cluster and took a swing at it. My thumb throbbed from the first blow it received. I improved, but I found the effectiveness decreased as the cluster's outer layers dwindled. Only accurate raps could finish the job. Clusters with three or four plump shells posed no problem, so I made quicker work of them. Smaller clumps with no pronounced central oyster proved far harder to whittle down. Jagged, tiny oysters, with likely no meat growing inside, shattered, became sharper, and refused to fully crumble from the impact of steel on shell. So I handed the bar and cluster back to Justin to separate the three good oysters left on the annoying bundle.

Justin noticed a circular oyster loose on the ground atop spent shells. It was a bivalve from his last farmed crop, likely dropped during a harvest. It was small, roughly two inches, about an inch shy of the length of the oysters we sought from the wild bed. Its edges were different, too. A lacey band ringed the oyster in a pattern of valleys and peaks. A translucent glow hummed from its tips. The shell itself seemed designed, with striations dark and light marking the widening animal as it enlarged from its anterior hinge. The wild oysters we had been culling were two toned: dark gray and darker. This farmed one looked tiger-like, with light and dark stripes of color. The enticing arrangement of orange and blue falsely made the farmed oyster look more organic and natural than its wild brethren.

Aquaculture made the difference, Manley explained. Oysters in the wild fought through multiple obstacles to survive to a harvestable size: blue crabs' claws cracked at oyster shells under the water; oystercatcher beaks pecked at them at low tide; the heavy settlement of new oyster larvae made those can-of-sardine-like cluster conditions that squeezed and choked out growth. Dense packing on the beds made oysters aim sunward, forging both the blade shape, which made the oyster a pain to shuck (since the hinge is narrow compared with that on wider oysters), and a shallow cup that housed a thin, long oyster. In mesh bags, tumbled by hand to loosen sediment from the shells, oysters advanced unencumbered by their neighbors. They ate to their stomachs' glee. They filtered phytoplankton and organic material from decomposed marsh plants through their mouths, directly into their stomachs, and then circled it through an intestine (all in the space of a square inch). The animal fattened by filter feeding. It expanded freely, growing deep in cup rather than long in shell. Growing in bags, the oyster could open its shell freely and eat in peace. There were no brothers and sisters to outperform, no need to grow long and slim to find a seat at the table.

Like marbling on an aged rib eye, a deep cup became a desired standard for half-shell oysters. Wild blade oysters were the chuck eye steak of the bivalve world. This lone farmed Spatking oyster we found perched on the oyster bed certainly was deep in cup, which was exactly to Justin's plan.

The deepest, most perfectly cupped oyster is the Kumamoto, and Justin's goal in crafting a farmed Georgia oyster was to emulate this Japanese breed. Largely unknown in its homeland now, Kumamotos bring home the bacon for Taylor Shellfish Farms, the largest aquaculture operation on the West Coast of the United States. Made plump over three to four years in culture, Kumamotos are bone gray with beautiful skeleton fingers forming a cup like fleshless phalanges. The taste is fruity, like cantaloupe, as Taylor Farms described it, a favorite on oyster bar menus across the country. A separate species from *Crassostrea gigas* (Pacific oysters), *Crassostrea sikamea*, the Kumamoto, was an inspiration in shape only for Justin. The specimen of Justin's we found freely tumbling about the oyster bed exemplified the traits: smaller, ridged, circular, but not perfectly round. Justin did not care to replicate the Kumamoto's flavor. Georgia oysters had their own idiosyncratic taste, one that he thought could and should be marketed to the benefit of himself and the state's watermen.

"Let's pop some open," Justin said, and we arranged a few oysters from our small gleaning on the skiff's bow, neatly in a line so we could examine their differences again. Classifying the shape of Georgia oysters was not easy. People call them coon oysters, which is a term used to talk about these weird, wild oysters up and down the East Coast and into the Gulf of Mexico. Others and myself have used the words *craggy, blade-like, clustered*, and *finger-like* to describe these bivalves, but other than *blade*, there is no clear defining word for its shape. Farmed oysters, the ones arriving on smatterings of chipped ice on silver serving trays, consistently look like commas or teardrops, wavering from this format only slightly and rarely. But these wild ones could be long and slim, or short and fat, or even boxy.

Justin grabbed a knife from the bucket and shucked a few oysters for us. He wedged the metal beak into the hinge, finding a hidden weak spot at the point where the two shells meet. He rotated the blade left and then right until its top half snapped open.

Inside is the animal, the oyster. By sight, I can't tell any difference in oyster meat from one location or another. Maybe Justin could; we never compared. But people like him looked at an oyster and told time.

A milky liquor filled the cup. This told Justin that it was ripe to spawn. The tissue itself was cloudy due to its reproductive state. The taste would be spawny, too. Justin warned, "They don't taste exactly as they should."

We don't eat wild harvested oysters in the summer because of spawning. It's illegal in Georgia to do so. The oysters need a break, and we eaters would be slurping a slightly off-tasting product anyway, so the five-month split is mutually beneficial. A slurping vacation is especially important in Georgia. Oyster science had shown that oysters spawn at least once a year, likely at the beginning or end of the season, or perhaps both. In Georgia, there are multiple spawns throughout the summer, a friskiness that exacerbates the clumping problem, which is called overset. Harvesting season ends on May 1 to ensure future generations of oysters; it isn't picky bureaucratic overreach, as Georgia fisherman who fought such restrictions on oysters and shrimp in the late 1800s and early 1900s believed.

Back on the mound, the cleft shell's meat had thinned out after its spurt, the portly muscle now gaunt. Postcoital remnants hovered in the oyster liquor, a briny juice that sustained the animal out of the water. When drained over our tongues, its saltiness purses lips.

The out-of-season mediocrity that Justin warned of did not stop me from eating these oysters. They were dirty, almost gritty, since they were unwashed. Droplets of mud kissed my lips, and some had crested the shell edge and into the liquor, adding a note of sulfur. The oyster's quintessential saltiness remained, and these Georgia bivalves oozed salt in ways some of their northern brethren can't muster. Bright, tingling, deep, like a mingling of humus and sea. Some oysters passed through the mouth without remark, like a cloudless sky, but this oyster, earthy and baked, lingered. I asked Justin his opinion of them: like lemongrass, he said, like pluff mud hardening under balmy air, spartina toasting under summer sun; a mouthful of the sea, intensely salty.

Talk of taste loosened Justin's tongue. He described the bivalves with vigor. But soon after, a burdensome pall returned to his face. It loomed heavier as we continued to discuss industry building. He had asked many people to come see what he was brewing deep in the marsh. Journalists wrote the story and moved on. They left Justin alone to accomplish what had been promised in interviews: a beautiful invigoration of a dead fishery. Justin held what he thought would keep that pledge in his hand. A meager length of plastic. We stood on this oyster mound, the marsh calm enough that we could hear a wick flicker, and he pecked with his fingernails at the chalky traces of spat on a stick.

The university, the government, the oystering community, and others longed to watch the industry find its way. But a shining example had yet to ignite. Watermen, regulators, and legislators wanted evidence. The Spatking and his stick hadn't yet provided enough.

POACHERS

A pair of boots pressing into marsh mud.

Justin grabbed an oar and pushed us off the mud bed where he had shucked a few oysters. We slowly motored back down the creek toward the Medway River.

Back at the dock, Justin hauled the boat out of the water alone. He roped the boat to the dock, ran to his truck, backed the trailer down the ramp, then ran back to the boat and powered it onto the trailer. Finally with an opportunity to be helpful, I hooked and tightened the boat to the trailer frame while Justin hopped in the truck cab.

He revved the engine. The vehicle backslid a few feet then found its power and hauled the *Green Hornet* up the ramp. Water washed off the trailer in gushes and continued to drip for yards as we pulled away from the Sunbury public dock.

Before Justin could return me to my car in Savannah, we had to drop off his boat back at Danny's place. Through a curving driveway guarded by a cattle gate, we lugged the skiff over pits and roots that jostled us. Eventually, Danny's homestead emerged behind a bend. A simple frame-up set on tall blocks, longer than it was wide. A door and a double-hung window on either side, with a pressure-treated staircase leading up. Like so many homesteads along the coast, a grid of leaves and branches high above kept direct sunlight out like a perforated umbrella. Light slipped through gaps in the dome. Humidity stuck our shirts to skin.

Boats, some seaworthy and some not, acted as traffic cones, and we swerved around and through them to back the boat into the right spot. I stepped down from the truck cab and saw crab traps, PVC pipes (from Danny's Justin-inspired spat stick experiments), and nautical flotsam strewn about the yard. I could see a neighbor's house, maybe two hundred yards away, reached by catapulting over brush and downed tree trunks. I swore I glimpsed a tidal creek shimmering through the woods behind Danny's house, like stars poking through a green filter. If what I saw was real, and it was as close as it seemed, the glinting scene hinted at how fast this lot would wash out in a storm.

Danny's girlfriend lived in Hinesville, and he spent his free time at her place. When at home, he lived alone except for the hogs.

"Mind those holes, man," Danny greeted us. The screen door smacked its frame when he exited the house. "Those wild hogs have been digging up the yard. They come through at night and just tear up the place."

Upturned plants served as proof of hog marauding, the black earth ground into piles like anthills by their late-night rooting.

"I can't wait to catch them," Danny said, resting a palm on a gun handle strapped to his belt. "I'll be eating good, too."

Dressed in blue jeans and a short-sleeved work shirt, untucked, Danny stood short and slim like a cowboy. Salt-and-pepper hair streamed out of a ball cap. We all stood around the bed of Justin's truck, and Danny, hands at his waist, looked as if the firearm pressured his right hip to bend.

Fifty-six at the time of our meeting, Danny was native to Liberty County. He had worked in construction most of his life with his brother Tommy, who died in 2010. But Danny was a born waterman, a true son of the marsh.

Under Justin's tutelage, Danny attempted aquaculture. He had collected spat, built trellises to hang bags of oysters. He had given it a shot, so I asked whether he thought oyster farming had longevity, and, just as I had asked Joe, what he thought tied up industry growth.

One answer to both questions: "Thieves."

Poachers, as thieves of oysters wild or farmed are often called, strike after dark. Guided by flashlights, they boat toward commercial leases and strip oyster beds of a few bushels and then sneak off into the night. Now that watermen like Danny were attempting aquaculture, local thieves found a new target. They had hit Danny hard, stealing everything bagged up and hanging on his lease. He never got a chance to sell a farmed oyster. It wasn't likely, Danny explained, that he'd ever see a return or that the criminals would be punished.

Around the marsh, poachers are rarely caught, even though the culprits are usually known. A DNR agent must catch them in the act, and accusations without evidence do very little.

An oyster poacher could be a waterman's neighbor, even a waterman himself. Once Joe Maley caught a crabber on one of his prominent oyster beds. The fellow had been stranded there and was waiting for high tide in order to escape with bushels of oysters that Joe had left. Joe sat and watched him through binoculars, made his presence known. As soon as he could, the thief skedaddled. But Joe had recognized him. Next time he saw him on the marsh, he stopped and took pictures of the crabber—not for any purpose other than to freak the guy out a little. That was about all the payback Joe would ever get.

Danny had a good idea who had robbed him. While he loudly threatened to show the culprit the business end of his shotgun, the same menace he wanted to offer these hogs, Danny knew it would never come to that.

The DNR kept tabs on a number of suspected thieves and brought charges when they could, but the biggest deterrent to poaching was punishing potential thieves by denying them a commercial shellfish lease of their own. The DNR did what it could to stop poaching, but it couldn't secure the safety of any oyster on any lease.

Poaching was a serious problem, not just for fishermen, but also as a public health concern. Consumers had to know they were getting a safe product. How would they know shellfish had been stored properly? How long had they been out of the water? With poached oysters, the DNR couldn't say.

Aquaculture would ease the way for thieves.

Danny put it bluntly: "We're setting a buffet for those assholes."

Danny's case was bad. But a few bushels here or there off a wild lease was nothing compared with oyster poaching around the world. In Georgia's marsh, catching thieves required having eyes on the criminals. In busier oyster grounds, the job increasingly fell to technology.

Along the Maryland and Virginia shorelines of the Chesapeake Bay, poaching had long been a hobby of down-on-their-luck fishermen looking for extra cash in lean times. Virginia's Marine Resources commissioner called poaching an "epidemic." In 2013, in part to protect the oystering boom, the State of Maryland turned on a radar system that acted as a virtual fence around protected oyster grounds. If a boat crossed the fence, a ping would notify an officer of the Maryland Natural Resources Protection Division. A patrol would be dispatched and the poachers apprehended.

In Australia, growers of the country's famous Sydney rock oysters turned to a piece of microtech employed mostly by the auto industry. An oysterman named Alan Burge, after losing $9,000, fitted his oysters with tiny metallic dots visible only under infrared light. If someone pilfered his oysters and the goods were recovered, he could identify his product.

Radar and microchips weren't coming to Georgia anytime soon. For Danny to get justice, he would have to park his boat up a creek, lay his shotgun across his lap, and wait until those jerks showed themselves after dark. But that was more effort than any waterman would give. Danny, like the rest of them, resigned himself to losing oysters.

"You spend your whole life doing this," he said. "You can't handle someone coming out and taking it."

We left Danny's compound behind and drove in silence back to the interstate. Something weighed on Justin. He had seen me scribbling notes while Danny railed against poachers.

"I don't want to influence the kind of thing you're going to write, and I don't want to tell you to not write anything you think you should, but . . . I guess I'm worried that with us trying to get all this going, I'd hate for anyone to get scared away."

Could I not mention poaching?

"I totally understand," I said, "but I can't make those kinds of promises."

"I get it, I do," he said, and seemed relieved, at least a little, and we changed the subject.

We talked about booze, our shared love of it.

"I can't drink whiskey anymore," he said. "It used to make me a monster. Back when I was lifting all the time, I could just rage. But man, do I love craft beer."

"Do you still lift?"

"I do, but it's just to keep me from feeling too old. Got to stay in shape."

"Have you ever tried to sell your oysters in Atlanta?"

"I tried to get something going with a wholesaler, but it didn't work out."

"Anybody up there try and find you?"

He pulled out his phone and began scrolling through emails, dangerously, with one hand on the steering wheel. He found it. An award-winning chef, a TV star with a small empire of restaurants, had searched him out. The chef emailed to ask: Could Justin UPS oysters to him in Atlanta, just like a few oyster farms in Virginia do?

"I admit, I hadn't thought about UPS," Justin said.

"Are you going to do it?"

"No," he said. "I can't."

"Why not?" I didn't understand why he wouldn't drop everything and get his oysters to the highest-profile chef in the state.

"I don't have any more. I'm done. I quit. I've already started selling off all my stuff."

This news struck hard. It made no sense.

He had told me himself: every oyster he grew, he sold. Who quits when they're selling product at such a high percentage?

"I can't make it work, not financially. I mean, I launched a start-up, so the costs were high. Whatever was coming in just barely caught me up."

Despite big sales, the Spatking had dipped too far into the red. "Too much going out, too little coming in." The culprit: gas, mostly, for that rumbling Detroit tank he loved dearly. Had he lived closer to the marsh, like Joe or Danny, maybe the books would have balanced.

Oysters like the single one he had plucked from the spat sticks garnered him the attention, but the way he went about creating and growing them was so time consuming that labor inputs overshadowed merchandise output. A hatchery definitely would have helped him out, but in the end, it really was just about the gas.

On top of all that, complaints about his lot in Thunderbolt kept coming, and it was getting worse; the city wanted him out. Get your boats and gear off the property, they demanded. Justin consulted a lawyer. The cost of fighting the city and community would be steep.

The pressure warped him. He hated all of it.

"I'd been getting it from every which way," he said.

Amelia's job supported the family. Their expenses clearly tilted towards Justin's corner of the ledger. Their savings propped up the Spatking. He felt selfish. His wife paid the bills while he was out splashing in the water, draining their bank accounts in pursuit of aquaculture dreams.

When he told Amelia, it didn't go well. I could imagine her body writhing like an air dancer balloon summoning dirty vehicles into a car wash. Maybe she slammed her fist down on their kitchen table, eyed him with her icy blues, and shouted, "No!" loud enough to wake Bella and Mercer. Their condo wouldn't have been big enough to absorb the sound.

Amelia loved oysters, loved the food scene, restaurants, loved that her man was part of that world. She loved oyster farming more than Justin did, he joked. Her license plate on her SUV proved all that: OYSTERS.

When it came down to fighting for the future of Spatking, Amelia taped up her boxing gloves. Justin had already thrown in the towel.

"When she protested, it only made me more resolute," Justin said.

He told her, "I can't do this to us."

I asked what he planned to do instead of oyster.

Get a job. He had already lined one up that put his scientific background to use for a company called Test America. He would be testing semivolatile organic compounds in a laboratory: air quality, chemical residues, that kind of stuff. Nine to five. Steady check. Boring as hell—that he knew. But there would be insurance, and his family needed those assurances.

So that was that. The Spatking was done. He had been done, but let me come down and meet him, spent a day toting me around the Liberty County marsh, all the same. He cared enough about the future of Georgia oysters that he had spent a day and a tank of gas to help sell the story to yet another journalist.

I had raised blinders to Justin's potential flaws, a pass I often afforded to people who worked with their hands. Backs aching, dirt under their nails, their toil often went unnoticed by the rest of us.

Maybe he needed a break, and he would eventually continue oystering. If that wasn't the case, his lecture about the future of the oyster industry in Georgia felt hollow. Was there any point in a hatchery?

Justin was a general leading a ragtag band of watermen recruits. They campaigned to storm the capitol. Now, he had abandoned his troops, and it made no sense.

Justin had a plan. Nothing firm.

He had witnessed the industry's faults and choke points firsthand. Before that, he had run the gauntlet of academic marine science. He had seen each world, knew the strengths and weaknesses. Justin figured he could harness his dual educations and continue the fight.

"I don't know how just yet, but I think I can be a greater help outside of the industry," he said.

This was why he had taken time with me despite, at the moment, getting nothing out of it. He had committed to the idea of oyster farming in Georgia. He was out for now, but definitely wasn't letting go of that dream.

In the meantime, Justin would have to dry dock the *Green Hornet*, recycle all that PVC pipe somehow. Give up the marsh and all its sounds and smells and breezes and gnats and marauding tiger sharks for a sterile laboratory.

Back at the Starbucks, I shook Justin's hand and thanked him for the tour. Please keep in touch, I asked. He would, he said, then he drove away. Past Thunderbolt and that lot that pissed many people off. Toward that condo on Whitemarsh Island where his wife, daughter, son, and all that financial guilt he had heaped upon himself waited.

He was going to adopt a new routine: wake up, get dressed in clothes not too fancy but far nicer than hip waders and T-shirts, get the kids to school, go to work, put on a lab coat, turn on his mass spectrometer, and start testing vials of water extracts.

Later, his gaze locked onto a gas chromatograph, he would stir every now and then and look up at the narrow rectangular windows at the top of the lab's exterior wall.

Through those small slits, light filtered through. He would see oak leaves shaking, boughs bending in rougher weather. On lunch breaks, he would step outside and fill his nostrils with fresh air. Then back to work, back inside.

Being penned up like this would be worth it only if someone, perhaps himself, eventually set him free.

OUT OF THE WATER

Newly formed oysters visible under a microscope.

Justin's career testing semivolatile organic compounds didn't last long.

Six months after we said our good-byes outside the Savannah coffee shop, he received an offer that hastened his exit from Test America. He respected his bosses there, respected the work, but the drudgery made him feel like a dog trying to fly. Justin preferred his old tricks, and now an old friend called with an out.

There was life in him yet, that he knew. Pride urged him to provide financially for his family, to not let Amelia stock the pantry and keep the lights on by herself. Oystering could never produce security, he was aware of that. But a life spent in sterile compound analysis broke his spirit with far more malice than the cobra-headed waves that had tried to drown him that stormy October day up Jones Hammock Creek.

Oystering gave him instincts he hoped he would never lose, but he couldn't deny they had dulled. By working in the testing lab, he had traded the visceral for the dependable. Dazed among the machines, the paint couldn't peel fast enough. Like an enclosed animal, he spent his days anesthetized, stifling screams that turned to moans that turned to whimpers. Get me out of here.

Amelia easily noticed the depression that Justin fell into. Lifting weights had long been a hobby of Justin's, one that in his youth had bordered on obsession. Early in their relationship, Amelia and Justin bonded over a shared love of working out. After the kids arrived, short gym routines kept the couple healthy amid the stress of parenting. So when Justin's gym bag gathered dust, she knew something was wrong.

After being outdoors daily for so long, Amelia knew that being stuck inside was intolerable for Justin.

"He retreated. He was negative. It was not a good time," Amelia said.

Late in 2013, Tom Bliss, director of the University of Georgia's Shellfish Research Laboratory, contacted Justin with a pitch. His office housed part of the Marine Extension Service, a version of what the agriculture departments of land grant universities offered dirt farmers. Their service transferred discoveries made by researchers to the real world, made laboratory findings applicable in the field. Extension agents read the methods, results, and discussion sections of scientific papers with titles like "Trends in recruitment and *Perkinsus marinus* parasitism in the eastern oyster," then figured out how to teach its lessons kinesthetically.

Tom had an agent position opening up. Was Justin interested in the job?

Back in grad school, Justin had witnessed the job description in action: meet shellfishermen out in the marsh and troubleshoot problems with their oysters; run experiments on fish populations and water quality for environmental agencies and nonprofits; collect and deliver old oysters shells for reef restoration projects. In short, the job promised time on the water. Not everyday, but definitely more opportunities than assessing the chemical structure of dirt samples allowed.

The job held other benefits, like decent pay and flexible hours that would allow him more time with his family. But even if it had been minimum wage and overtime, he would have justified a way to accept it. Incentives wouldn't lure Justin; he just needed to be back in oysters.

Tom still had more in store.

Back when Justin still oystered and maintained a role in the Georgia Shellfish Growers Association, the watermen had asked Tom to explore ways to bring oyster aquaculture to Georgia. Tom struck out on a few grant proposals before finally making a connection. With the help of the Coastal Resources Division of the Department of Natural Resources, Tom applied for a grant to fund a start-up hatchery at the shellfish lab.

"Increasing eastern oyster aquaculture production in Georgia through establishment of an in-state oyster hatchery" was the official title Tom gave the project.

If the plan met with federal approval, and it looked promising that it would, a pilot project to promote oyster aquaculture in Georgia would kick off the following spring. The grant funded a hatchery manager position, someone who would raise baby oysters and nurture them until they were healthy and big enough for open water. Justin, Tom told me, was uniquely qualified for the position.

The hatchery wouldn't immediately begin pumping out oysters. Its manager would have to build it, and initial funding had been allocated to do so. The plan for the first year, as Tom envisioned it, required collecting wild spat on sticks, just as Justin had done commercially.

From wild stocks, the hatchery manager would create a line of lab-raised Georgia oysters. Melding nature with hard science: nobody could do it better than the Spatking.

Do the extension agent thing for a year, Tom suggested. Get the project started, build the hatchery, then become manager and leave the extension work behind.

A no-brainer. It paid less than Test America, but Amelia told Justin not to worry about that. "Money isn't everything," she told him. "Especially when you're so unhappy." He started at the lab in March 2014, and Amelia noticed an immediate change in attitude. He rekindled friendships. He returned to the gym. The constant stress that had loomed over their home lifted. Amelia had her old Justin back.

Justin's new job was set: grow oysters, teach watermen how to farm them, rebuild an industry. Nobody could deny the tall order of the tasks. Before Justin stood before the assembled GSGA members in June 2014 as calmly as he could, the workload was theoretical, just words on a grant application and visions in Tom and Justin's imaginations. A room full of veteran oystermen now stared them down, wondering how the University of Georgia planned to change their lives. No turning back now.

The watermen who drove all the way to the shellfish laboratory had an idea of what to expect from the meeting. Charlie Phillips, Dan DeGuire, Joe Maley, Rafe Rivers, and Earnest McIntosh now awaited the official pitch. Free oysters and free farming gear, that part had piqued their attention. In the few years that Justin had farmed oysters, they had heard all about the Spatking's methods. For some, it seemed like too much work. But free stuff? Well, that was a good enough reason to try something new.

"So is this going to be like what you were doing with the spat sticks?" Dan DeGuire inquired.

"Yes, essentially," Justin said. "For this first year, until we have the hatchery going, I'll be collecting wild spat out there"—he pointed to a spot a hundred or more yards off the Skidaway bluff—"with PVC, no cement this time."

Justin referred to old experiments he had conducted on leases Dan worked. In a precursor to the spat stick, Dan had helped Justin coat PVC pipes in cement slurry and set them in the water. They hoped cement would increase oyster set. It did, but not as efficiently as they would have liked.

"What about panty hose? Can you collect spat on panty hose?" Dan asked.

"That, actually, is a really good idea," Justin said. It sounded strange, but it was possible to catch spat on panty hose if it had been cast with cement. "I haven't been very lucky stretching it out, but I'm not going to tell you how that happened."

Everyone laughed.

Before they went any further in particulars, Justin recognized the presence of Dom Guadagnoli and thanked him for attending. He then added: "Maybe Joe wants to say a few words."

Joe pivoted in his chair to face the group.

"Well, a few of us went to the Coastal Day up in Atlanta, and we shucked a lot of oysters. We had a really good response. There was one guy, he wanted to know how the problems with the Chattahoochee were bothering us . . . Jesus. Anyway, we'll be going again next year, in January, if anyone else wants to come along."

Tom stood up and laid out the project's time line. In the fall, participating shellfishermen would receive twenty thousand small oysters, he said. By next summer, there would be spawn in the hatchery.

"This is not just to produce seed," Tom told the watermen, "but to get it in your hands" and make sure those hands possess the tools and training to keep the seed alive.

Oystermen had to take ownership of the program, Justin said, because he and Tom needed metrics. Not just on oyster growth rates, mortality, and other bits of biology. But dollars and cents. How many did they sell? What price did they get wholesale for the farmed oysters? What about retail? Where did they sell them? Numbers, economic data points.

"We need to be able to map a market footprint," he said. With a detailed report on the economic impact of the hatchery, Justin and Tom believed they could press for the project's permanence, secure funding for the long term, and, importantly, keep a baseline of spat available until the industry became self-sufficient.

Tom reminded everyone about the decision to not allow out-of-state oyster seed imports. At a previous GSGA meeting when Justin was still a member, shellfishermen voted to not allow it, which made Dom happy.

"It wasn't an easy decision," Dom told me months later. "But I think it was the right one."

Justin wanted an endemic line of oysters for the hatchery, but there were other reasons to support the ban. Importing seed was a safety concern, Dom and Justin told me later. It was impossible to be sure the incoming seed was completely free of protozoan and viral pathogens that might threaten oyster populations. Why take the risk?

South Carolina watermen, Tom told the group, recently bought a batch of seed from Virginia tainted with a parasite—*Haplosporidium nelsoni*—that threatened the state's wild oyster population. In response, South Carolina placed a moratorium on seed imports.

"They're now looking at what we're doing," Tom said, meaning the creation of a line of endemic spat from wild populations, "and thinking it looks like a pretty good idea."

The project goal, Tom continued, is to hand off commercial oyster production to a private party.

"This place was designed to be a university hatchery and extension agency," Justin added. "It'll just be a research facility once it goes commercial. We'll just concentrate on breeding."

Dom then raised a few points. As the industry considered aquaculture like this, he said, it would have to go back to how the code was written. What was farmed and what was wild would have to be defined.

"The letter of the law speaks on behalf of wild oysters, not farmed," he said.

Someone asked about cultch. Rules stated that oystermen had to return bags of spent oyster shells back into the marsh, a percentage of what they harvested. The returned shells were called cultch, which also referred to pretty much anything on which an oyster successfully set itself. If these oystermen wouldn't be taking anything wild out of the marsh, would they still have to return cultch?

"The cultch law was put in there to ensure a sustainable resource," Dom said. "With wild and farmed, those lines need to be well defined, then we can seek reform."

"I think we should avoid using the word *farmed*," Dan said. "People might have a bad connection with it, like with Vietnamese shrimp." He had a point. Georgia's wild shrimp industry had a strong history that had been threatened by Southeast Asian imports.

"These are things we need to talk about," Dom said.

The second stage of the grant project involved buying and distributing aquaculture gear to the growers. But the kind of gear that might be deployed was as of yet undetermined.

Some in the room hoped for floating cages, buoyant racks that were accessible at any point in the tide. They had read about them in fishery magazines, had heard that they had been very successful in all sorts of marine environments and that they could potentially free oystermen from tidal tyranny. If they could work the water at high tide, their jobs would be plenty easier.

"It's our job to evaluate the gear and turn over our recommendations to the state," Tom said. Those evaluations would come quickly, Justin promised, but how the results would be used was up to the state. DNR would have to approve any gear for use.

Dom reminded everyone of the marsh's many stakeholders. Whatever farms watermen set up out there, they had to respect the needs of sportfishermen, too. Nothing built or placed in a river could be a navigation hazard. He preferred it if gear didn't rise more than three feet off the water bottom.

Erosion was another concern. Dom didn't want anymore anthropogenic wreckage of the marsh. He would have to wait and see what Tom and Justin recommended to him. Above all, he wanted to retain control, ensure he wasn't creating loopholes that allowed giant corporate aquafarms to descend on the marsh. Those corporations were already knocking at the door. To prevent their arrival, he cautioned patience with all things aquaculture. "I want to keep the footprint small," he told me. "It's moving faster than I'm able to keep up with."

"We'll be testing all sorts of methods," Justin said about the gear. "There's no reason for you to purchase the resources and then find out it doesn't work."

All the questions gave the room an energy that turned to urgency and excitement.

Joe Maley wished the grant process, and the results that came from it, could come quicker. We need to attract new growers, he worried. "Are there any leaseholders here who are under fifty?" he asked

Only one. Rafe Rivers, who had replaced Justin as the whippersnapper. At thirty, Rafe was close in age to Earnest McIntosh Jr., twenty-eight, the only other young oysterman in the room. Earnest worked for his father, enjoyed it, and seemed poised to take over once Earnest Sr., sixty-three, retired. Everyone wondered what changes might have come to the industry by the time any of them retired.

Dick Roberts sounded encouraged. "There are big things happening with oysters," he said. "The whole industry is starting to focus on the East Coast. For us, there's nowhere to go but up."

Justin asked for a show of hands. Who was interested in participating in the project? We'll need your commitment for two years, he said, and you can sell every oyster that survives.

Unanimous agreement.

"What's to lose?" I heard later in the parking lot. "Might as well give it a shot."

"Well," Joe said, ready to close the meeting. "I move for adjournment and benediction."

Justin and I watched the last of the watermen motor away from the shellfish lab. Their taillights disappeared into the darkness, and we were alone. Justin spoke of the honor of being back in the watermen's presence, of the confidence he had in the hatchery project. It gave him great pride to be helping the men he respected.

The Justin I encountered this evening differed greatly from the one I had met a year earlier. A guard had been let down, one of toughness, just as he raised another. Throughout the night, he chewed on technical words and spat scientific concepts that spun my head in circles. I understood him to be a smart, well-educated man, but he brandished jargon like a knife thrower unconcerned that his blade might knick my earlobe in his display of skill.

The ease with which he spoke this other language pleased him. He rattled off aquaculture argot at increasing speeds, unbothered by my glazed-over face, until I asked him to slow down. Yet he still marveled at the raw power of watermen, an ability he himself once possessed. Was he a scientist or a waterman? The weight of nostalgia that the departing oystermen's trucks brought upon him convinced me that both were viable personalities. One just seemed to be flickering a bit brighter at the moment.

Justin was a man of both romance and data. In the past year, he had moved from the marsh to a lab to yet another lab, one governed by a university's giant bureaucracy. In such a sprawling organization, what role would the once-autonomous Spatking find? For now, he was unconcerned with internal struggle. Only the mission mattered. At the shellfish lab, he became part of a tradition.

"What we're doing here is a step, built on the previous work of the last guy," Justin told me, thinking of his mentor Randy Walker. "I've read everything the [extension] service has produced. Randy has a body of research that has no comparison."

Justin considered the lab sacred ground, Randy a saintly figure, and now he had returned to the fold. The move gave him purpose.

"I'm glad to be in oysters again," he said.

DOWN AT CHARLIE'S

Captain Charlie Phillips in the offices of Sapelo Sea Farms.

The next morning, Justin and I went to meet "the man."

There was no operation on the coast like Charlie Phillips's Sapelo Sea Farms, Justin explained with an air of veneration. Georgia's oldest and largest clam farm. Respect was due. He was the source of some hard feelings, sure, but nobody had accomplished what Charlie had. Therefore, nobody wielded power like he did.

Charlie kept nearly a dozen McIntosh County residents on the payroll. And with a half-dozen leases under his watch, his staff tended to more square inches of the marsh than any other waterman. Charlie could flex some muscle and make oyster aquaculture viable, Justin said. Whether he possessed the will was another matter.

About a forty-minute drive south of Sunbury, Sapelo Sea Farms sat at the end a road marked, in progression, by mobile homes, aging single-family dwellings, retirees' homes announced by surname-emblazoned signs, and shrimping docks—small at first, then larger. On their postal addresses, these homes used Townsend, about fifteen miles inland, as the city name. But Crescent was the nearest community. The homes clustered along a Sapelo River bluff seemed confined within their own unique boundary, not just a nameless unincorporated community, but a thriving society apart from its neighbors. But every intersection here carried the hazy sense that you had entered a secret fiefdom camouflaged by a shared zip code. Languorousness was a secret weapon to lull outsiders into vulnerability.

Two marina-like edifices—one pink, one white—edged the bluff. This was Charlie's compound. Their construction appeared unplanned, with continued additions affixed at odd angles to an original box shape. Space upon space upon space in cobbled-together perpetuity.

The pink one especially caught my eye: multiple stories capped by a windowed tower that functioned like an air traffic control room, its height providing a wide marsh vantage; coral clapboard wrapped around a giant walk-in cooler; ladders that led to staircases that led to nowhere; stacked bags of freshly harvested clams laid on cement, sprayed down by young men dressed in half-peeled-off wetsuits, their shoulders sun toasted.

The men hauled in bags of clams from Charlie's many leases. Clams grew quite easily in the marsh. They could be laid out in long beds on the mud, since clams preferred a life beneath a layer of muck. When an order needed to be filled, one of Charlie's teams would roll up a sheet of clam bags and bring them back to shore for cleaning and sorting.

Not quite watermen, Charlie's workers were another endemic species. Regional, but wilder. Rednecks, perhaps. Country, for sure. Black and white. They were apt to make bad decisions, Charlie told me. They rush-jobbed their way through life, paycheck to no paycheck to garnished wages for child support.

"I bet half of my people are on probation, are in jail, or been in jail. They're good folks, most of them, most of them just have issues. One of them called me up: 'I won't be in today.' What happened? 'I was playing poker and I got shot. But the good news is, it's a small caliber, so I'll be back in a few days.'"

"That's the stuff I get," said Charlie, dressed in jeans and an orange shirt, his hair shaggy and his shoulders slouched.

Across a gravel drive stood another patchwork building. This one was a brick ranch with wooden extensions. Inside was the sea farm's office, staffed by Charlie's sister and two or three others. They answered telephones and crammed papers into overloaded filing cabinets.

The room was dark and air-conditioned, a relief from the outdoors. Photographs of employees past and present lined the walls alongside maps, framed licenses from governmental organizations, invoices, alerts from the National Oceanic and Atmospheric Organization, smart-alecky posters about cleaning up after oneself, directed at employees from management.

We shuffled into his office. Sun hats, wetsuits, sunglasses, and water sandals mixed with paper stacks, an old IBM, and rolled-up maps. After a bit of rearranging, we all found seats.

Charlie leaned back in his chair. A ball cap covered bushy white hair, and his voice whistled out from under a mustache like a harmonica.

"We didn't draw a paycheck the first couple of years we were in this business," Charlie said.

Sixty at the time of this meeting, Charlie was born in Jesup, about forty-five minutes west of where we sat. There, Charlie's dad, Myron Phillips, owned a chicken feed mill. When Charlie was in the seventh grade, they moved toward the coast. An entrepreneur, his father bought a shrimp boat. Charlie passed his teenage summers as a deckhand. By the time he graduated from high school, in 1973, he was spending his weekends as a boat captain.

He tried college, working days and studying nights, going for an associate's degree in psychology. But one day in 1974, Charlie decided on an alternate path and quit school. He told his professor, "I'm not going to be here tomorrow. I've got a boat to run."

Then he told his dad. "I made sure I was out in the yard when I told him," Charlie said.

With financing help from an uncle, he bought his first boat. He paid that one off, and in 1980, he bought his second. He hired a crew, which included, at one time, Jeff Erickson and Mike Townsend, and worked from Georgia, around Florida, and into the Gulf of Mexico.

In 1990, Charlie bought the pink shrimp-processing plant that visually intrigued me outside. His dad purchased a restaurant that still operates up the hill from Charlie's office.

He started wholesaling shrimp from local fishermen. When shrimp imports increased, he tried his luck with deeper-water fish such as tile and snapper.

He credited his success to his flexibility, which he said was necessary because of shifting governmental regulations and environmental factors. The big catch today won't be there tomorrow, or the government won't let you go get it, his theory went.

"You just keep digging, keep morphing," Charlie said. He planned to keep up his fishing boat addiction. The money he made from clams, he joked, kept feeding it. According to a financial disclosure statement he filed when he became a member of the South Atlantic Fishery Management Council, Captain Charlie Phillips owned three vessels that hunted snapper, grouper, king mackerel, Spanish mackerel, etc.: the FV *Sea Puppy*, the FV *Fish Hound*, and the FV *Capt. Lynn*.

A worker at Sapelo Sea Farms washing clams after a harvest.

On top of all this, Charlie was looking at blood arks (a type of bivalve) and sunrays as his next seafood dalliances. He predicted a pretty good profit from each.

"But if the water quality goes bad, it doesn't matter," he said. "It's all dead."

With Charlie, there was always a bright side.

Charlie's interest in clams began in the late 1990s. Randy Walker and Dorsett Hurley, two marine scientists, had attempted to introduce clam farming as an economic engine for a community of Gullah Geechee, ancestors of enslaved Africans who live in relative solitude on Sapelo Island. They struck out. No interest from the island. When they scheduled training sessions, only one old man far too frail to do the work attended, Randy told me.

Sapelo was a no-go, but Randy and Dorsett still had all the farming gear and plenty of clam seed to dole out.

"Randy wanted somebody who'll pay attention to what they were doing, keep track of things," Charlie recalled. He and Roger DeWitt partnered up to buy an island near Brunswick. It became the first testing ground for the burgeoning industry. "The first thing we learned how to do was kill clams."

In a few years, Charlie bought out Roger. Given Charlie's shrimp and fish wholesaling, the infrastructure he needed to make clam farming successful was already in place. He possessed a network of distributors and customers who were able to carry clams from his Sapelo River dock to Atlanta, St. Louis, and even Canada.

Those early years weren't easy for Charlie and the other clam farmers. The Department of Natural Resources had matched UGA's investment in the project by staffing the new industry with a field biologist, a program manager, a marine technician, and a laboratory technician.

By the time Charlie jumped into the fray in 1997, most of those positions had been eliminated. And in 1998, an essential function of the DNR's shellfish program—the kind of water-quality monitoring necessary to meet FDA requirements for commercial shellfish sales—became vulnerable to budget cuts.

The reasoning: the shellfish industry didn't seem to be expanding, even after pushes by the university and the state. Sales had stagnated at $100,000. There wasn't much point, the DNR argued, in keeping it alive.

Charlie counted himself among furious watermen.

"Everything we've made has gone right back into building this business," he told the Morris News Service in 1998. "[Sapelo Sea Farms] has no value except for shellfish. It is worthless, absolutely worthless."

He and other watermen fretted that if water testing halted, even temporarily, the industry would be ruined: the FDA would institute a three-year ban because of the lapse.

They pressured the DNR to include funds for testing with the funds for public health monitoring along the state's beaches in the 1999 budget.

The clammers' public outcry averted a crisis.

Charlie became the face of clam farming in Georgia, just as Justin had been for oysters. Charlie used his minor pulpit to protect the environment. In newspaper stories related to commercial fishing, his comments tended to follow the lead. He often donated his clams and wild oysters for roasts that benefited river keepers and conservancy organizations. When we met in his office, the two biggest issues riling him up at the moment were the threats of offshore drilling, which was promoted by Georgia's U.S. Senate delegation, and a new interpretation of marsh buffer law, which had kept residential and commercial development far away from the high-water mark but might lose its teeth.

A handful of environmental menaces scared oystermen like Charlie and conservationists alike. A Texas energy conglomerate called Kinder Morgan was attempting to bully through an oil pipeline that skirted the eastern edges of the estuary. It planned to obtain necessary land via eminent domain. If usurping land didn't upset enough people, the potential for crude leaks did. Oil pipelines were faulty, and marshlands and all their feathered and shell-based stakeholders likely wouldn't recover from the slightest mistake, pipeline opponents warned.

In Camden County, where Dan DeGuire and Dick Roberts lived, economic development efforts had attracted a spaceport project. Heavens-going spacecraft would be launched from a bluff site overlooking the marsh. Nobody quite knew whether the launches could poison the ecosystem, but riverkeepers I spoke with worried over the blast of development—housing and ancillary businesses—that would surely follow such a project.

In Charlie's hometown of Jesup, a private landfill sneaked through permits to clear land near the town dump to house coal ash deposits. With coal ash came arsenic and its plant-killing, water-quality-ruining power. The rural people of Wayne County, a little upriver from Charlie's marsh but still out of earshot of the Atlanta economic and political elite, were having none of it.

Spaceports and pipelines and buffer repeals had two serious foes on the coast. The first, a piece of legislation. In the late 1960s, a conglomerate of scientists and politicians passed the Coastal Marshlands Protection Act. Their efforts came in response to a mining operation's bid to dredge the top layers of marsh mud and spartina, then dig for phosphorus. The act ushered in decades of slow and modest development along the coast, keeping small towns like Sunbury and Eulonia from overstretching the natural balance between man and marsh.

The second, the people of the coast: sportfishermen, marine scientists, cordgrass philosophers, kayakers, hunters, small-town journalists, lawyers, cantankerous hoots, and shellfishermen like Charlie.

If he was to put it simply, without waging a "Stop the Pipeline" campaign or anything, Charlie just wanted people to give a crap about what they threw away, sprayed on their lawn, and ate.

"People don't care, don't understand, don't know," he said. "If they knew, they might try to mitigate it, use less nitrogen, or be more particular about where they put concentrated feedlots."

Speaking of feedlots, did we want lunch?

Charlie, it turned out, was particular about food. He avoided GMOs and ate mostly fresh fish, stuff he assured us contained no pollution.

"I promised the banker that I'd eat healthy until I paid his ass," he said.

We retreated by truck to Charlie's house, about one quarter mile away, where he planned to make us snapper tacos. We met Trouble, his dog.

"He's ornery, I don't know where he gets it from," Charlie said.

"It's the salt in the air," Justin said.

With Charlie's life story out of the way, and tacos consumed, we broached oysters. Charlie, it turned out, was pretty skeptical. He hardly messed with wild harvest himself, opting to buy from the McIntosh family, the Timmons, and Jeff Erickson, Mike Townsend's brother. He was a clam farmer, and raising another type of shellfish seemed to me to be less apples and oranges than oranges and tangelos. But for Charlie, the numbers didn't add up yet.

"I sell fifteen thousand clams a week," Charlie told us. "I'll have to sell three hundred bags at seventy-five dollars each just to get everybody paid and keep the lights on. You can't get nickel-and-dimed to death. Now, the price we can get for oysters is higher, but we don't know what kind of yield we're going to get."

The bags are more expensive than clam ones, he added. All the variables haven't been run through.

"It's going to be a lot of work," he said. "I can't be trifling with something that won't make me money."

Just like that, his tone took a positive turn, albeit briefly.

"You have to go about it like it's going to work. Oysters have a good shot at making it," he said. "I don't think the taste for oysters is going away."

Then a joke, sort of: "Or, someone could get sick off eating an oyster, and it would all go away. But I guess, even then, people have short memories."

If oysters took off, would he consider giving up clams?

"I'm not going to leave the girl who brought me to the prom," Charlie answered.

No naïveté in Charlie, not a drop. The wide eyes that Justin saw the future of oysters with weren't shared by the captain. By the end of my time with Charlie, I could tell he exercised a fairly accurate ability to assess an economic opportunity for risk and reward. And that assessment was his intellectual property, not yours. Hooded, drooping eyes kept shrouded whatever it was he actually thought. His smile and wisecracks belied the strategies forming behind wireframe glasses.

Once oyster aquaculture proved itself feasible here, I sensed that Charlie would throw himself into the task with a fervor unmatched within the shellfishing community. For now, he would wait and see. At least that was what he wanted us to think, and that was the impression we left with.

"I think I learned as much as you today," Justin said to me as we left Charlie behind at Sapelo Sea Farms. "I really never think to ask about the history of people."

Justin glowed from all the waterman repartee; he had been saturated with Charlie's folksy wisdom. His infatuation with watermen, I could see, might never fade.

"Some of these guys didn't graduate high school, and they don't consider themselves educated," Justin said. "But I look up to some of them like rock stars."

JOE AND LESTER

Joe Maley.

All this talk about how the Georgia oyster industry would reinvent itself, but what, exactly, would be changing? Justin had attempted to show me a typical waterman's labor when we had visited his oyster farm, but the example didn't fully satisfy. Aquaculture promised to deliver a cleaner, prettier oyster to urban eaters, but what impact would it have on oystermen? I had been told that there wasn't much difference between how Georgia oysters were gathered today and thirty years ago. Except for the invention of the outboard motor, the methods might not have been altered in a century. Joe Maley offered to take me back in time.

One winter day, Joe let me hitch a ride as he and a helper gathered a few bushels of wild oysters. It would be a long shift, he warned. We would start late in the morning and not return until well after dark. But he promised a good time and some lunch.

Joe's lease wasn't too far from Justin's. We would head out the Medway River, the same route taken to reach the Spatking's oyster lease. We would be going further, though, into St. Catherine's Sound, which gathered the waters of the Medway, North Newport, and Bear Rivers and flushed them out into the Atlantic Ocean. In the sound, we would steer north, keeping clear of a long bar called the Medway Spit, and scoot northwest up the Bear River.

Joining us on our excursion, Joe informed me, was a shrimper named Lester. In the off-season, Lester, who had relocated to Georgia from North Carolina not too long ago, picked up work here and there on Joe's boat. Joe had railed against the poor labor stock available to him in Liberty County, but he could depend on Lester. Lester, though, preferred shrimping: its weeks at sea were followed by stretches at home. So it wasn't too often that Lester answered when Joe called.

Lester arrived outfitted in sweatpants and a sweatshirt. Roughly the same height as Joe, he bore wide shoulders and a belly over ample thighs, a build common among older laborers of all sorts. A knit cap almost covered a large welt over his eye, the source of which I refused to inquire about, given Lester's mum demeanor. Unlike Joe, Lester tendered no folksy quips about the authenticity of fishing. He asked for a day's work and the night off. His philosophy could be summed as getting it done. Lester's reticence didn't quiet the trip. Joe talked a boatload.

Joe's constant companion came in the shape of a circus poodle named Bobbie. On marsh errands, the curly-haired butterscotch dog trailed along wherever Joe went, mostly silent but always skittish.

In the morning, Joe, Bobbie, Lester, and I met at a private dock that extends from the Sunbury Crab Company, a restaurant owned by Joe's brother, Barney. The restaurant served Joe's oysters—steamed or on the half shell—when in season, as well as crabs caught by Joe's nephew, Clay.

Joe moored his vessel there in fancy company: ropes kept a few sail boats and vacation cruisers bobbing next to his Lindsey Charter boat, an old commercial tuna ship Joe bought off a longline fisherman in North Carolina. The boat was built to travel forty miles out to sea. Joe never took it past the barrier islands.

"It's not to be confused with a pleasure boat, though being on it can be pleasurable," Joe said with a wry smile.

Boats hadn't always been pleasurable for Joe. Like Justin, Joe didn't grow up on the water. It had become an obsession as an adult. The first few decades of his life took him from Alabama to southern Georgia farm country to California to Germany to Seattle and to Middle Georgia. When he finally landed in Hinesville to work at Fort Stewart, being on a boat became his one and only hobby. He befriended Danny Eller, who knew little other than his career working off the marsh. Joe began joining Danny on weekends to harvest oysters or toss a net for shrimp. Danny always operated the boat, Joe remembered, so he had never had a reason to learn to run a seagoing vessel. But when Danny got injured one day, Joe had to learn on the fly.

"One time [we were] pulling in the [shrimp] nets. We were rolling them over the gunnel, and something in there, a stingray or maybe it was a catfish, got Danny in the finger, and he about passed out. His finger was turning black, and it was up to me to get him back to the hill.

"The water out by St. Catherines [Island] is blue, but there's a lot of bars, and I had to drive us back. I tried driving, and he's yelling at me like, 'Go over here, head over there.' I'm just hitting them all. So he just ended up driving the boat back himself, one-handed. That's when I knew I needed to learn how to operate."

So he learned, in his forties, to captain a boat. But that didn't mean his troubles had ended. He still had to figure out how to read the weather.

One of his first boats was a yellow John Dory, the same one that was upturned in his yard the day Justin and I paid him a visit.

"A lot of folks were scared to get in it because of the low sides, but it was very seaworthy," Joe said. Until it wasn't.

"It was a nice sunny day, calm. I had all my [oyster] bags on the boat, getting ready to come back around [the Medway] Spit. I started back, and I noticed it was pretty choppy. So I stopped and secured everything in the back before I was going to tackle that. About that time, a Coast Guard chopper comes in and hovers above me. There were gale force winds coming. Maybe they were trying to warn me or just looking at a stupid old . . .

"I came around that corner, and it was like Malibu surf crashing up against the back. It was cold that day, and I was taking water over the sides. I knew something had to change or I would die. It doesn't have to be freezing to get hypothermia and die.

"I turned up the first creek I saw, and assessed the situation. I had a radio, but nobody could pick up. I didn't know what to do: hike up a bank and go build a fire?

"Wind was from the south to southwest. It was five- to six-foot waves. I couldn't take the river back.

"I decided to go into it, cross the sound, and it was a booger. A lot of acts of contrition on the way over there. But I was able to get back through to Sunbury through the creeks.

"That's why I have the boat I do now. It can take water."

The Lindsey's hull was deep, the gunnels rising above my knee, and it had plenty of room for the orange baskets in which Joe gathered oysters. Its engine, housed in a fiberglass belly, had until yesterday leaked oil. Joe had spent the day before zipping back and forth to a parts dealer, hurrying to fix the vessel because he had oyster orders to fill and the boat couldn't be grounded for long.

Lester hopped in the boat and lifted the shell off the motor as Joe carefully lowered his body into the hull, letting out an achy groan in the process. Joe started the motor, and the two men watched the engine come to life, shaking and fuming, looking for spurts of grease.

I waited at the stern in a plastic chair, cornered there by Bobbie, whose trust I had not yet gained.

All good, Joe nodded toward Lester. After loading a cooler filled with ham-and-cheese sandwiches, donuts for "fortification," waters, and sodas, we were off.

As Lester operated, Joe shifted about the boat, stacking oyster baskets and, finally, picking up Bobbie. He sat down next to me on the rail as the boat slowly motored through the Sunbury Channel's no-wake zone.

Joe waved his hand to draw attention to the Sunbury bluff as we passed by. Dormers on new construction peeked out between trees. As we moved away from dry land, homes became harder to see. It became the "hill," as waterman refer to the upland, a dense wall of green, civilization creeping somewhere behind it.

"All this was in play 250 years ago," Joe said. What we could still see of the hill had once been a plantation, much of it privately held from the colonial era. He wondered why Europeans came here in the first place, because Sunbury's current residents had a love-hate relationship with the coastal climate: "The heat, the humidity, the fucking bugs."

Hardly hot this January day, cloudless for sure. A breeze zapping the air of moisture. It was sweater weather, at worst, and no sign of gnats yet.

We picked up speed through the Medway and then came through St. Catherine's Sound, around the spit, and finally up the Bear River. The winter sun to our left—the west—and Ossabaw Island off to our right.

The dash in front of the steering wheel where Lester sat looked like that of a busy contractor: papers strewn about, crushed soda cans stacked liked boulders, gloves missing one of the pair. A typical sight when a vehicle serves as an office or a second home.

Lester navigated us toward an oyster mound called the Head.

Eleven years ago, Joe took over a shellfish lease almost barren of oysters. In his first year, he pulled maybe a few bushels off it. He knew what he was getting into by becoming its guardian; part of his stewardship of the lease meant rebuilding its bivalve stocks.

The Head, Joe felt, proved he had accomplished that task. He turned a mound of mostly dead shells into his most productive bed. By moving shells from old middens onto the Head, he expanded the base on which oyster spat could settle.

Some of his mound-building methods were less organic. Underneath healthy layers of oysters sat junk, all sorts of it—mailboxes, two-by-fours, old metal fencing—anything he thought oysters might want to settle on. A flat cobbled-together version of Justin's spat stick.

"Some days I came out here looking like Sanford and Sons," Joe said: a boat laden with odd trash and an oysterman tossing bits of it into the river at planned intervals. Joe clearly loved this quip. The line leaked past his lips syrupy slow. He added extra twang for effect and rubbed his belly like it was the source of his wit and wisdom. It was hard not to like him.

With the tide still mostly out, we couldn't see the full extent of the oysters living in the Head. A small mound, maybe four feet across, rose out of the glassy water. Lester motored past the Head and doubled back toward it.

Closer now, Joe took out a wooden rod and poked it into the water, gauging depth.

"This'll do," Joe said, picking up an anchor off the deck and handing it to Lester.

Lester leaned over the side of the Lindsey and let the steel anchor swing two feet below his hand. He accelerated the arc he'd created, eyeing a target. With one big heave, he launched the anchor into the air and it landed—plop!—some distance from the visible Head.

"Now we wait," Joe said. "Once you're here, you ain't going nowhere. We're stuck until the tide comes back in."

Any trip to the Head required arriving well before it could be seen. It made scheduling a workday a chore for Joe—never quite knowing when he would be heading back home. Although it did have its pluses.

The Head spent most of the day hidden like a wall safe. Most people would never know it existed. But there was one flaw: for a few hours every tide cycle, the safe was left unlocked. Yet nobody could rob it.

High tide cloaked the Head's bivalve treasures. At the high-water mark, no oysters could be seen. Potential poachers would boat by oblivious. At low tide, not even a canoe could reach the stash. A sweep of mud separated the oyster mound from the nearest navigable water. Impassable.

Joe opened the cooler and offered sandwiches and drinks. Lester snagged a donut and changed clothes, ditching tennis shoes for boots and pulling up waterproof bibs over sweatpants. Joe felt fine in a hoodie and sneakers. In jeans and rubber boots, wrapped in a windbreaker and topped by a skullcap, I was somewhere between the two. We would be out here all day, and comfort in the elements mattered.

We had an hour to kill. And time to chat. Lester was skeptical of me and my camera, taking pictures of him driving a boat and tossing an anchor.

Lester throwing the anchor off the Lindsey into the Bear River.

"We should take a picture of the cameraman," Lester shouted at Joe. "Get your phone out."

Lester scowling, with Bobbie in his lap. Me on the rail, mouth full of ham and cheese. Cute picture. Joe later told me that Lester took me for a secret agent working for the DNR. Somehow, snapping a picture of my agent self was akin to capturing a ghost.

The tide dropped by a foot, and it was time to position the boat. Still anchored, Lester brought the Lindsey around so that the bow faced the Head. With a shot from the engine, he rammed the vessel into an unseen part of the mound. Shells crackled and the hull moaned as it rubbed against oysters. The boat slipped backward.

"Not quite. Try one more time," Joe said. With a little more force, Lester thrust the boat forward. This time it stuck. The Lindsey wouldn't be moving for a few hours, not until high tide. Water still lapped at the hull, so we had to wait a bit longer.

Lester spoke of sharks in the river. I figured he was trying to scare me. Joe joined in.

"Somebody saw a baby tiger shark down in McIntosh, maybe five feet long," Joe said.

"Here, look at this," Lester said, pulling out a smartphone. He found a photo of a shark jaw a friend had found out in the marsh. An immature specimen, perhaps, but still big enough to swallow my head whole.

"They're out there, and you never know," Lester said. Impression made.

As we changed, ate, and found new sources of fear, the river disappeared and revealed the full breadth of the Head: as long as a shotgun house and twice as wide, so extensive that the receding tide marooned the Lindsey on an island of mud and shell. Two mounds encompassed much of its acreage. In parallel to a nearby marsh bank, the Head's boundaries extended north and south, the edges fanning out season by season as Joe added layers of old shell onto soft mud. Each application of shell broadened the blanket on which oysters could set. It wasn't a precise science, like adding a couple rows of corn to an already fertile field, but it worked.

Maintenance of the Head never stopped. Luckily for Joe, it was a profitable chore. The Head was his honey hole. Oysters grew well here. The Bear River's deep tides meant lots of wave action over the crop, strengthening and shaping their shells. It seemed counterintuitive to crash a boat into Joe's cash cow, but he wasn't concerned. The oysters could take it.

The pressure of a boat's beam presented little danger to the shell-encompassed animal. Hell, we were about to walk all over their bodies, and it wouldn't faze them. Joe and Lester planned to deliver blow after blow with rigid harvesting tools, and the oysters would survive the impact.

"Where's our knockers at, Joe?" Lester said, looking through the cluttered dash and the hull cubbies packed with wet suits, life preservers, and oars. After a search, he found them. Lester's knocker: a flat piece of steel, about a foot long. Joe's choice? A painter's friend, one of those five-in-one hand tools with a scraping edge that spreads putty and cleans brushes, a point to open paints cans and open beers. Each fit easily in their back pockets.

Joe grabbed a five-gallon bucket; Lester took two. The ship's bow was now fully out of the water. To get down, Joe set up a ladder. Otherwise, it was a six-foot jump, maybe more. Both Joe and Lester's bodies ached too much at night to attempt something that brash. Slowly, they descended.

Lester breaking down oyster clusters on the Head.

Two main mounds made up the Head. The biggest one currently cradled the Lindsey. A gully defined it against another lump of oysters. From the two masses, a shag of shells unrolled, allowing us to walk widely from the boat.

Taking advantage of the spread, Lester marched off as far away from Joe and I as he could get. He found a perch that looked south down the river, upturned one of the buckets and popped a squat. Legs stretched wide, he leaned over and plucked a bouquet of oysters from the mound and took a few whacks at it with the bar. Shell chippings and mud flew, splattering near the tide line. His Coke can nestled perfectly between two clumps of oysters.

A trio of willets approached Lester, close but keeping a safe distance. Their beaks pecked at the mud, looking for invertebrates. Bobbie saw them, gave the sandpipers little yips. They didn't interest Lester much. He slung gobs of mud in their direction.

Lester and Joe both referred to the wild cluster oysters they were harvesting as "coons." This being the South, a land fraught with racial tension, I wondered, the first time I heard the term used, whether "coon oyster" was connected with the derogatory, Sambo-esque caricature of enslaved Africans as "coons," popularized in the media by characters like Stepin Fetchit.

I had heard black watermen, Lester being one of them, use the term, but that didn't mean its etymology wasn't racist.

My hunch was incorrect. Two reasons for the "coon" term existed. One, that the long and thin shells resembled a raccoon's paw. Two, that raccoons feasted on cluster oysters whenever possible. Old newspaper clippings gave evidence of both theories.

Lester held up a large coon specimen for me to inspect. He had chipped down a cluster until one large oyster with two smaller ones attached near its hinge was all that was left.

"That's a pretty good looking one," he said. "It could be a roaster or a single."

I heard Joe yell from the other side of the Head.

"What's the matter with that oyster, Lester?"

"It ain't in the bucket!"

"That's right."

Joe smiled widely. Stomping around the territory that he had struggled to make viable and that had rewarded him in return, he relaxed and talked jovially.

"I mean, you can't ask for a more perfect day than this. The water is just like glass, the sun is high, and there's a breeze and no gnats."

After a while, Lester opened up as much as he would. The two men batted about a few lewd jokes and talked about strip clubs. But in a show of age, when they assessed nudie bars, it was about the food: "That one near Jacksonville, man, it has the best hot wings!" Lester said.

Joe smirked, working away with his painter's friend at a clump of oysters. He sometimes used a railroad spike, the dense metal easily prying oysters apart. All he had today was the five-in-one. "This [plastic] handle is real bad for leverage," he complained.

Lester and Joe plugged away, lugging buckets full of oysters back to the boat, tossing the contents into bigger buckets sitting in the hull. Back in the bird-free safety of the Lindsey, Bobbie watched from a chair. They hoped to fill three of the large buckets before dark. They were getting close.

The light changed hue as the hours passed. Everything—the men, the spartina field in the distance, the Ossabaw tree line—seemed cast in a blond film, as if the air itself sparkled with gold. In these last silent stretches, with Joe and Lester's heads bowed and focused on their toil, the scene resembled a painting by Jean-François Millet. I saw the men as if the French artist had secured the unhurried but serious grace of this gleaning on a canvas. Nostalgia for what I was witnessing set in. The incoming wave of aquaculture threatened to turn this livelihood into an anachronism. Despite his role as a player in the changes being set to speed, Joe sounded as if he hadn't yet decided exactly how he felt about it all.

"Justin said farming will be easier work. We'll make more money. Maybe that's so," Joe said. His face and sweater were covered in splotches of gray mud. "Like I said, this is labor intensive any way around it."

That was why he was out here, I realized: he liked the work. He had a comrade in Lester, but I could assume that Joe liked it out here all alone just fine. Just Joe and his thoughts.

"Not everybody gets to have an office like this," he said.

To be honest, there were downsides: the elements were fierce, just not today. But I would still get to sample a drawback to oystering.

As the sun drew down, the breeze, our only defense against bugs, lessened. The gnats attacked. Necks, cheeks, foreheads: anything uncovered felt the wrath. It bothered me—them nipping and me slapping myself—and bothered the tough guys, too. I wondered how much longer we would be able to stand it.

An unseen bird cackled. To Joe, the calls made by marsh hens hiding in the cordgrass were a sign.

"You know what they're saying?" Joe said. "We ain't heard them all day. They want to go to bed, and it's time for us to leave."

Dusk had officially arrived, and we packed up the boat with the last full buckets. The tide flowed back, covering the outer offerings of the Head's smorgasbord, and came faster with each wave.

Eventually, with enough water under the Lindsey, we shoved off, using rods and oars to push us clear into the deeper river.

Silence occupied the return voyage. At full night, with tired bodies, there was little to say.

The marsh in twilight seemed an unknowable expanse, an abyss worth staring into, but too infinite to utter words within its presence.

The Medway in darkness shone like a silver stream, and we glided along its argent path. Moon rays glinted off the curling spartina, our unseen wake lapped at the levee marsh, lost in blackness.

Joe switched on a dim light above the wheel. It alone guided us back toward the spare streetlights of Sunbury, where by now most of the town's residents had returned from jobs in far larger places, cleaned up after dinner, and nestled themselves into couches.

Joe and Lester's workday wasn't over: the oysters needed to be cleaned and bagged in bushels. Another hour yet, and then they would be home.

Not being done for the night didn't stop Lester from thinking about how he would spend the $100 coming to him. Finally in cell range, Lester scrolled through web pages on his phone. He searched wholesale steak prices. The glow lit a serious face intent on finding a deal on dinner.

I thought back to the morning when Justin and I had first gathered around Joe's coffee table. Joe complained about labor: "The sons of bitches got to show up at 6:00 a.m." At least this day, one of those sons showed up to work for Joe, and kept him company.

At this hour, exhausted and distracted, Joe and Lester could at least count on each other, no matter the hunger level. If something had gone wrong, Lester would have backed up Joe. Vice versa. If all hell broke loose, they both had Bobbie.

In Justin's Spatking days, he had always worked solo. The headstrong quest to rebuild an oyster industry by himself was difficult, and Justin behaved as if he could attain his goal only through monastic solitude. He never intended to be entirely alone on the water. He hoped others would witness his methodical approach and imitate or join in. He wanted people to stop saying it couldn't be done. All the talk of dirty little coon oysters bugged him.

The isolation had defeated him. After spending a day with Joe and Lester, I couldn't help but wonder whether Justin left the water, finally, in search of companionship. That was true, to an extent, but it wasn't another human he sought.

At Skidaway, Justin found millions of friends in the form of microscopic oysters. They couldn't survive without him. He was needed. That feeling refreshed him. What was more, he learned he couldn't survive without them. Larvae would become medicine, healing him one spawn at a time.

"They were more my genesis than me theirs."

THE LONE WOLF

Mike Townsend.

Justin spent the summer of 2014 growing oysters in the Skidaway River, and everything went well. Summer for oystermen was an off-season, a time for fixing boats and tending to crab pots. For a former oysterman like Justin, summer's extended sunlight could stretch on forever, but he still needed another minute to finish his tasks. Prolific oyster spat filled the water column, and it was Justin's job to catch as much of it as he could. He tried to nab every last floater in the bunch.

Through June, July, and August, as planned, he collected spat from the PVC sticks he had stuck upright in the Skidaway River, popped off the nascent oysters, and stuck them in protective bags.

From the size of a pinhead, the oysters grew, some larger than others. By the time I visited Justin again in the fall, with plans to explore McIntosh County with him and visit watermen there, some oysters had plumped to a quarter inch in size, some to a half inch, and some a bit bigger than that.

As he collected spat that summer, Justin also built a version of the upweller that he had described to me before the GSGA meeting in June. Aquaculture people called it a "flupsy," a floating upweller. Justin used pressure-treated wood to construct the flupsy frame, a platform that was kept buoyant by air-filled barrels. Eight square boxes that Justin called silos were attached to the frame, each one filled with oysters sitting atop a mesh bottom. Water flowed up through the boxes from the river, then out through a hole near the top of the box. A propeller pump controlled the flow of water over the oysters, delivering microscopic algae for the developing bivalves and removing their waste.

Ropes kept the flupsy tied to the Skidaway dock. Water splashed within the silos like waves against a hull, and the flupsy bobbed in the river as fishing boats cut wakes.

Until the shellfish lab was complete, Justin's makeshift hatchery was completely outdoors. He had disinfected, plumbed, and readied the lab to receive the fiberglass tanks that would store oysters through their pre-marsh stages, but delivery day was some time off. For now, the flupsy and an aluminum table he employed in oyster sorting constituted his production facility.

Tens of thousands of oysters passed through the flupsy that summer, and some had been sent out into the marsh to be cared for by watermen. When I peered down into the silos, Justin told me that he figured he had already pulled between 350,000 and 500,000 baby oysters off the spat sticks.

He set aside a few of the big ones, oysters with deep cups and quick growth rates, to serve as brood stock when he got the hatchery online. He would spawn those indoors and collect sperm and eggs for future batches.

A good number of oysters had died. Once a week, Justin lugged a silo onto the dock and dumped the contents onto the sorting table. Under a typical Skidaway sky—a shocking azure sheet devoid of shade clouds but with a rain blast threatening from the west—and accompanied by an ensemble of boat motors and shorebird mews, Justin culled lifeless bivalves from the bunch. Their tiny gaping maws signified their death.

Plenty of them lived.

Justin was able to set aside 200,000 for watermen, roughly 20,000 per lease. In the week before I arrived, Justin had dropped off parcels of oysters, but hadn't reached every participant yet. Early response from the guys sounded positive, Justin said.

"I think before, they didn't know what they were getting, it's all been abstract," Justin said. "Then they saw the oysters, and were, like, 'Holy shit, this is mine. I'm going to make some money.'"

Justin hoped that with the GSGA guys trying to farm all at the same time, they would share a sense of camaraderie by all doing something new in tandem. This might have been too much to ask, since the guys saw and spoke to each other rarely. If so, Justin would settle for helping them make extra scratch.

"I want them to realize they can make money, that they can chase their own dreams," he said. Justin estimated, understanding that not every one of the 20,000 seed would survive, that watermen might earn an extra $7,000 from the project. He hoped it was enough cash incentive to keep them involved.

When he distributed seed, Justin often handed them off, exchanged a few words, and left. The watermen did with the packages as they pleased.

He might offer advice on how to construct their farm, but only if they asked for it. Justin couldn't be too protective of the seed he grew; watermen wouldn't respond well to overbearing instructions. They would adapt this novel approach in their own idiosyncratic ways. It was best for Justin to lie back and let the guys figure it all out for themselves.

Justin trusted the inherent ingenuity of watermen. They were a smart, fix-it-yourself breed of men. If it all collapsed, Justin hoped they would seek his aid. But he wasn't worried. Anyway, he had to keep looking forward to the hatchery. He had to anticipate hiccups in the project and predict what needs watermen might have once this first phase found success.

While the old salts didn't need Justin, the new guy did.

Rafe Rivers, the organic farmer, acquired a fifty-acre lease over the summer that had been willingly carved from one of Mike Townsend's river plots. Rafe knew nothing of oysters, and wild harvest held no interest for him. Farming was his specialty, and he was eager to apply his trade to shellfish.

Rafe knew the minds of chefs and knew how easily he could sell a beautiful farmed local oyster to his fine-dining customers. Problem was, at this point he could hardly find his lease, much less set up a farm. Luckily, Rafe's background in sustainable agriculture meant that requests for technical assistance were a matter of routine. In Rafe's world, farmers solved problems together and never hesitated in calling their taxpayer-funded extension agent. Pride never entered the equation.

When Rafe asked Justin for help, the extension agent jumped at the opportunity. The plan was for Justin to drive Rafe's seed allotment down to his farm, called Canewater, in the community of Carnigan, just a few miles north of Darien, the county seat. Together, Justin and Rafe would ride out to his lease and assemble trellises on which to hang bags of oysters.

On the way to Rafe's, Justin planned to make a stop near Crescent and hand over seed to Mike Townsend, the lone wolf of the marsh. I must meet Mike, Justin told me. He was a character, and Justin considered him the quintessential waterman, a case study in the type's resiliency and ethos. With limited means, Mike had turned clam farming into a viable enterprise. Justin admired how he had made his living over the years, and counted on him applying that same pluck to oyster farming.

Justin recalled his first encounter with Mike: a skinny, compact guy, maybe five-six and 150 pounds soaking wet. He slung eighty-pound bags of clams around like a shot-putter. Justin watched Mike's feats of strength, the size that belied the man's power, and winced at how little Mike charged for clams and oysters.

"He was underselling himself," Justin said. He hoped to stop such practices by Mike and his colleagues. This initial shipment of seed oysters was a first step in achieving that aim.

Justin loaded pails of seed for Rafe and Mike, and farming media destined for Rafe's new farm, in the white diesel Ford truck he now operated.

"Shouldn't take us too long [at Mike's]," Justin said as we exited the shellfish lab parking lot. "We have to be at Rafe's by noon at the latest if we're going catch the tide."

Interstate 95 South. Forty-five minutes. The Eulonia exit. East toward the water.

On a state highway, we approached an intersection anchored by a popular country restaurant, a gas station, and a pharmacy. From the traffic light, a left turn would take us to historic downtown Eulonia, now a collection of small houses, a Piggly Wiggly, and a bed and breakfast. A right turn would send us to Darien. Behind us, a post office and the county health department.

We stayed straight through the blinking red. Buildings became scarce. What nonresidential buildings I saw had been long scuttled, boarded up. Quick stops sold beer and snacks. There were front-yard car detail shops, shade tree mechanics, but not much else.

As we neared Mike's place, Justin placed a quick call to announce our imminent arrival. One minute later, Justin turned right down a gravel drive marked by a mailbox and warnings: "No Trespassing" and "Beware of Dog." If we had stayed on the main road, we would have hit Charlie's in a few miles, Justin reminded me. We curved around one wide tree and then another. Behind a screen of brush as tall as Justin's truck, I could see Mike's land. Under the roof of a pole barn sat the makings for construction projects. A female figurehead carving dangled by rope from the barn's rafters. I counted the number of boats in the yard (five) and wagered that the total seaworthy square footage outpaced Mike's dry quarters.

Walking down deck stairs, Mike greeted us with a wave. He wore a white T-shirt tucked into unbelted jeans that were themselves tucked into rubber shrimper boots. The bangs and neck ends of his salt-and-pepper hair appeared self-cut, jagged like a forest of stumps.

Justin Manley and Mike Townsend.

"Y'all care for a glass of tea?" Mike said, his words stretched out, the impact delayed, like watching a frog lash its fly-wicking tongue in slow motion. "You need sugar?"

Mike lived alone in a stick-framed house that he had moved into the previous year. The embedded greenery and weathered outbuildings that surrounded the house—clad in aluminum—contradicted its newness. The house was the last piece of a puzzle: it completed Mike's tiny kingdom, a place of pride and worthy of defense. The property represented a long-term investment. For eighteen years, Mike had lived in a trailer here. He had bought it for $2,000 and paid next to no taxes, which allowed him to rack up savings. This new house increased his taxes exponentially. But since he paid for it all in cash, he remained indebted to no one, just as he liked it.

The inside of Mike's house was spare and meticulous compared with the museum of marine vessels and paraphernalia he had curated around the yard. Both were clean scenes, but the oxidizing menagerie outside and the dusted, vacuumed interior did not match. Most fishermen lived like this: their yards function as workspaces, and their homes as tidy refuges.

When he walked inside to fetch the glass of tea, Mike took off his white rubber boots upon entrance. He never wore shoes in the house. "Cuts down on cleaning," he said. "I only have to vacuum every six months."

He told me not to bother taking mine off. Stay there, he said. Sugar? he asked again. As he pulled a reused Styrofoam cup from a stack, telling me that they were clean and that he didn't have cooties, I scanned the room: nautical charts piled onto a coffee table; dining table stacked with a commercial fisherman's licenses and permits. The room brimmed with the paper trail of his work life, but kept a sense of organization, despite the fact that a strong breeze could destroy the filing system.

We caught Mike on a rare day off the water. He had interrupted his aquatic duties for a spell, but kept busy with errands and truck repairs. Taking a day off was a way of proving that even his beloved marsh, the place that helped him make a living for nearly four decades, couldn't control him: "I ain't fishing today because I didn't want to go today," he said.

Maybe he would enjoy a bit of relaxation and watch some public television, but that wasn't likely. On land, Mike was out of his element. Rest was fleeting.

He could handle three days like this, off his boat, off the marsh. Maybe he could stretch it to four, but only in horrible weather. "I can't sit on the hill too long," Mike said. "If I don't go by the third day, I just go to take the boat out, circle around the islands, check things out, get off the hill."

To Mike, the hill meant dry land. Up there, on the hill, people could knock on his door, and he had no use for visitors. Up there, on the hill, you had to talk to people, and Mike wanted none of that.

"I'm rural," he said. "I don't like people knowing what I'm doing. I'm low-key."

Being on the water was as off the radar as he could get. His reluctance to seek out others explained the moat of oaks we had waded through to reach his house and the warning signs nailed into tree trunks. Watermen were quiet, I had already learned, and reluctant to share words with outsiders. Despite having two outsiders in his realm, Mike still respected the rule book of southern hospitality, which explained the offer of tea.

Mike was fiercely independent, a loner who refused to work under the pressure of another man's thumb. His stubborn quest for autonomy had cost him friendships, he told me, and probably cost him financial gain. But he didn't care. He owed nothing to nobody.

From this attitude, the rift between Mike and Charlie Phillips had originally arisen. Mike and an early partner—a coastal bigwig named Roger DeWitt, who let Mike harvest clams off his Crown grant—had had an opportunity to become Georgia's first clam farmers, with the guidance of the University of Georgia. They would be given gear and a new lease. DeWitt asked Mike whether he wanted to go into the clam-farming business with him. "Hell, yeah," Mike remembered saying. "It was free!"

As the new venture's launch day approached, Mike's partner said he was adding a third member to the team: Charlie Phillips.

"I saw real quick what was happening," Mike said. "They were going to be making money off my labor. I wasn't going to have any of that."

So Mike respectfully bowed out of the relationship, not ruffling feathers, stating that he preferred to work alone. He quit harvesting wild clams off DeWitt's Crown grant and made plans to apply for his own shellfish lease. With some effort and help from a colleague of Randy Walker's, Mike kicked off his own clam-farming operation. It was far smaller than the one Charlie was now running alone following DeWitt's retirement.

Mike continued to sell clams to Charlie, who sold wholesale seafood around the lower forty-eight, but never trusted him. Charlie was yet another potential usurper of his liberty, trying to make a living off other people's living, Mike thought, and he considered suspect anyone unlike himself—unbought and unbossed.

When I talked with Mike in 2014, he told me about a date from nine years earlier that he had memorized: April 5, 2005. The last time he sold Charlie clams. He had had enough. Enough of the price haggling. Enough of that guy's attitude.

Mike didn't need anybody else selling his clams for him. That was just another thing he could do himself. He didn't depend on anybody else. Only two things mattered: his freedom and the water.

"If you can't go do what you want to do, it ain't worth shit," Mike said. And what he wanted to do was simple, merely the same thing he had been doing since he could remember. Fishing. Splitting Doboy Sound into big wakes with a sputtering outboard motor. Letting the sun crisp his skin like an oven broiler. Meditating in immense silence of the marsh like a monk. Out there, all alone. The only small talk to avoid was him reckoning with his inner voices, his harshest critics. That he could handle. It was where he was born to be.

Since dropping out of school to join a shrimping crew at sixteen, Mike had worked only two weeks on the hill—two terrible weeks at a gas station. He earned 140 stupid measly dollars.

"I hated it," he said. "Worst two weeks I ever had."

With Mike, it felt safe to employ a cliché I often heard about watermen: he had salt water in his veins. It felt safe because I knew I had met a person for whom the cliché might have been composed. I squelched an urge to grab the knife from Mike's belt, poke him in the abdomen, and see whether brine spurted out.

That the marsh could wield terrible power did not deter Mike from a life among its storm surges. Mike knew well that humans on boats tempted fate. On June 6, 1964, his father and grandfather, two factory workers, drowned while sportfishing in Doboy Sound, near the Duplin River at the south end of Sapelo Island. Mike was only three. When Mike reached his teens, his uncles introduced him to water life. Despite the fatal warning proffered by his elders, Mike committed himself to the wetlands with greater vigor than they had.

As we talked, he referred to the countdown until he could draw social security. Seven to ten years from now, he would be done, he said. "I'm going to find a sailboat, and I'll be in the Bahamas. Need to find a woman to take care of me." But there was no way he would give up this life. He told me he expected this lunge at oyster farming to be his last project, but I didn't believe that. When he talked of retirement, that moment on the horizon represented yet another shedding of expectations. He could collect a government check, and that meant he wouldn't have to clam or oyster if he didn't feel like it, which meant he didn't have to sell any of us a clam or oyster if he didn't feel like it. "I love my job and I love what I do," he said, softening a bit. "And I really shouldn't be so 'Oh, I'm hanging it up.' I love my job, but it's what I do," he repeated. "It's just me, all alone."

Mike's hermitic tack gave me the sense that while Justin had ridden from Skidaway to McIntosh County with a truck full of oysters to give Mike, free of charge, it was Mike who felt he was doing the favors here.

Justin lowered the tailgate, and he and Mike began sorting oysters. They dumped a bucket of oysters sized between a quarter inch and a half inch onto a steel table and counted them to let Mike roughly gauge what

Sorting oysters.

the preferred allocation per bag should look like. Justin fired off some instructions about what locations suited these oysters best for grow-out. Flats. Low silt load. Other jargon I missed. But Mike understood it all.

"You mean up a creek and propped up where they won't get a bunch of mud on them?" Mike said. Exactly.

"Remember, this is not a handout," Justin reiterated to Mike, knowing how to stir the waterman's conservative and miserly morals. "You are helping us build an industry. Once we show the state"—and here he meant DNR, which might change regulations to promote oyster farming if it liked what it saw—"that you know how to use this stuff, that it works, they'll let use more efficient equipment. We're trying to get this to where the labor cost is doable."

"We're going to have to get a premium price, too," Mike said, which launched the watermen into a tirade about seafood distributors, the kinds of corporations that underpaid the "independent, family-owned farms" that represented "small-town America." People like him, that was who he meant, people who always got cut out of making real money.

He railed against the major fresh-fish distributors in Georgia, the companies responsible for delivering to restaurants whole fish caught as close as the Atlantic and the Gulf, clams, and oysters from the increasing number of domestic aquaculture operations.

Mike contended that the prices distributors were willing to pay for local oysters would put small-timers like himself out of business. How do we sell enough to make a difference without involving the big guys? That was what Mike wanted to know.

Red embers seemed to char Justin's cheeks. His temper rose as this discussion heated up. Behind his sunglasses, I imagined his pupils becoming as focused as lasers. Mike had worked him up.

"Here's the thing, Mike," Justin said pointedly. He didn't like all of Mike's negativity, that I could tell. Look forward, think positive. "We're going to have to play them like they play us. Undercut them. Work around them."

"You mean like the co-op?" Mike said.

"You bet your ass!" Justin said.

We had to go, Justin told Mike. Call me if you need anything, he added. Conversation over. Mike, used to riling himself and others up like a boiling teapot, shook our hands and waved us away while he walked back into his lair. Back on the road, Justin still fumed as we raced to Rafe's.

"Mike is absolutely the best waterman I've ever seen," Justin said. He tried to squelch the frustration caused by the micropolitics involved in increasing the meager wealth of men like Mike.

"What was that about a co-op?" I asked.

Back when Justin was in business, he and a few oystermen had formed a co-op, through the GSGA, that sought to circumvent distributors' perceived stranglehold on the industry. They might not yet be able to get perfect, farmed product to big cities like Atlanta, but maybe, if they arranged large shipments of wild oysters and did so dependably, Georgia oystermen could make a mark upstate.

All this depended on a refrigerated truck. And Mike owned the truck.

The way Justin told it, on the eve of the co-op launch, Mike backed out.

The way Mike told it, his participation in the arrangement relied on two major clam clients in Atlanta keeping up their side of an agreement. They backed out, and Mike sold the truck. Done deal.

But Justin still believed his watermen could "break the yoke of the powers that control the shellfish industry." Accomplishing such a rebellious feat required farming, Justin argued. "Oysters are the only way to do it," he said. Nothing else is as luxurious and highly priced.

But it meant somebody had to buy a truck. It also meant tamping down watermen's introverted tendencies. It might also mean, Justin was beginning to realize, that not every waterman could break the chains, imagined or real. Nor did they want to.

But maybe, just maybe, this wasn't at all about yokes, chains, or any sort of binding. Perhaps this wasn't about freedom, but about a new future and a new breed. The players in this future might not look anything like Mike, his brother Jeff, or even Joe Maley.

To Justin, it had become increasingly clear, especially after a run-in with Mike, that the next golden age of oystering in Georgia required the particular traits and skill sets of one type of person. So far, only one leaseholder possessed these attributes. For better or worse, he was a prototype, untested. He needed Justin's help.

MUD RIVER

Rafe Rivers walking through newly sown rows of sweet potatoes on his Canewater Farm in McIntosh County.

Eight miles of sorghum-slow road bends later, we entered the community of Carnigan. Looking later on a map, I saw that two of this community's streets were spelled Carneghan and Carnochan, each an iteration of the names of the area's Scotch-Irish settlers. When it came to naming the river that separated this dryland place from the marsh, spellings shifted, depending on the map consulted.

At a stop sign in front of a Baptist church, Rafe waited for us with the tail of his farm truck facing the highway. He would lead us through the transition from paved road to dirt path, eventually to his land.

Justin slowed his truck as we approached the left turn, rolled his window down to wave at Rafe, who then pulled away. We passed mobile homes and, seen through thickets of brush, other structures of indiscernible make and upkeep.

We opened and closed two farm gates that Rafe kept locked to ward off intruders. This being McIntosh County, with its reputation for political corruption, bootlegging, and general wildness, as chronicled in the book *Praying for Sheetrock*, by Melissa Fay Greene, fences prevented nothing. Deer fencing that ringed Rafe's farm kept out only night munchers, not night marauders.

"We've been seeing and hearing things at night," Rafe said, explaining the fences. He lived alone with wife, Ansley, and they expected a daughter soon. Glints of flashlight and what certainly sounded like rifle fire, these were not sights and sounds the young couple wished to abide in the dark.

Soil upturned to grow peppers, eggplants, and tomatoes, even this far into fall, looked black and dry, like compost flecked with sand, despite no lack of moisture. This fertile humus was not the compacted red clay that Georgia was famous for.

Rafe, thirty-one at the time of our visit, hadn't been on this land for long, although he had already made an impact. Kale sprouts poked out of the ground near the season's last round of nightshades. In just over a year, he had cleared four fields for planting, with plenty more in the works.

This level of agronomic diligence wasn't what I expected to find hiding behind a border of blue-collar houses, in a county without a clear economic engine. But just as Justin's entrepreneurial spunk set him apart from his fellow watermen, Rafe's time clock was not set to coastal time.

With Rafe still in the lead, we skirted his homestead: two rustic cabins, sided in weathered lumber, attached by a breezeway; an in-law suite, similarly clad, where Rafe and Ansley kept their offices. Rafe circled through a clearing, weaving around oak trees, and backed up his farm truck at a dock. Justin followed suit.

Rafe had carved time out of his usual farm day to spend the afternoon on the water with Justin, and his outfit showed he hadn't yet left the dirt behind. He wore brown duck denim pants, flecked with grease and dirt, tucked into muck boots, and a blue long-sleeved sweat-wicking shirt. He kept his long ponytail—a rich brown color with thin gray stripes—tied in a tight bun at the base of his neck.

He strode around the truck to greet us with confidence, a man for whom each movement's purpose had been weighed beforehand. Energy for labor had to be conserved. Waste not, want not.

While Rafe's physical presence impressed, nervousness and worry bothered him, especially over this oystering business. Although he knew so little about it, he desperately wanted to be part of it, but was scared, and hoped Justin could fix that. It didn't start well.

Justin unlatched the tailgate and slid the spat to the edge.

"Do you have the bags?" Justin asked. Rafe pointed to a pile of mesh bags, the ends closed with zip ties. Justin inspected them.

"These aren't going to work," Justin said. "The way you have the bottom folded over on the outside like this, it's just going to let crabs in and let oysters slip out. They're too loose, too."

The mesh bags opened on each end. Rafe's chore was to close one end in preparation of being packed with seed. He had messed it up. Rafe laughed, embarrassed.

"I was up all night doing that," he said. His voice sounded soft but daubed by dust. The delicate grit hinted at hoarseness from yelling across fields to farmhands.

Justin pulled out a pocketknife and sliced through the zip ties.

"I've got plenty more," Justin said, grabbing a bundle of them from his truck.

He called Rafe over to show him the seed. Justin dug into a bag laden with pebble-sized oysters. He grabbed a handful, and tiny shells, twinkling like glass, cascaded over his palm.

"Now these are too small, you're going to just hang these off your dock today. The rest we'll plant," Justin said. Together they fastened the bags, tight and secure for the future farm.

Justin had also tasked Rafe with preparing the rebar and PVC pipes, cut into eight- and four-foot lengths, respectively, that would form a trellis. He had also drilled wide holes into one end of the PVC.

It was past noon, now, and to catch the lowest tide, we would have to leave soon. There was no chance we would be back before dusk, but we still had to hurry. Rafe's lease was a forty-five-minute boat ride away.

Today, Justin expected a neap tide, an exceptionally high one, which meant the Mud River, the location of Rafe's fifty acres, wouldn't fully drain. Justin and Rafe would be arranging the oyster farm in water at least up to their knees. As the tide washed back in, that height would reach their thighs, eventually their chests.

With Rafe's two-person fishing boat laden with steel, plastic, oysters, and three humans, we cast off from the dock. We slowly motored through Crum Creek, named after a cantankerous hermit who once ruled the area, past the tabby ruins of the Thicket, Georgia's first rum distillery, now just a foundation for cedar trees and live oaks. A few private docks swayed in the afternoon glare.

Once the creek met the Carnigan River, Rafe pushed the engine into high gear. Wind and splashes of salt water whipped our faces. Justin directed from a seat in front of the boat's wheel, telling Rafe when to bank and which side of the river cut the deepest channel.

I observed Justin, a man at peace, in a way, in this moment. His gaze was fixed on marsh layers: sweet tea waters, impenetrably dark; erect blades of spartina, a gradient of verdancy to translucent yellow; the sky, an unmarred canvas.

For a day on the water, he had dressed in basketball trunks and a T-shirt with a fleece vest on top. He wore tennis shoes.

An oyster shucked with a flathead screwdriver.

Black crud had invaded the creases of skin on his hands, remnants of degreasing the lab and changing the oil of boat motors. One hand held a sixteen-ounce can of Monster, an energy drink that kept Justin mobile on long days like this. His right leg fidgeted from caffeine and excitement; Justin was at his happiest.

"I couldn't sleep last night, I was so excited for today," he told me. Then he craned his head to yell at Rafe.

"Have you thought about a name for your oysters?" Justin asked him.

"We were thinking Canewater, after the farm, or maybe Crum Creek," Rafe said. "Maybe Mud River makes more sense?"

"You'll have to figure out who you are," Justin said. "I was the Spatking, that's what I sold. You sell who you are because you know who you are."

"Don't compete with other oystermen," Justin continued. "Be your location and sell it. A Doboy Sound oyster versus an Ossabaw oyster is as different as pinot grigio and pinot noir, and that's what you sell."

That sounded hyperbolic to me—perhaps because Justin shouted this advice over the whoosh of wind against our faces and the drone of motorized propeller—but I liked the sentiment. In Rafe, an organic farmer catering to high-end restaurants, Justin found a receptive audience for his merroir marketing spiel.

"Listen to you, you are like a motivational speaker," Rafe said.

The Doboy Sound, a wide body that separated Sapelo and Wolf Islands, interrupted the Carnigan River. While we were in that open water, waves rocked our boat and swifter winds pinched exposed skin. Soon, we were back in the relative protection of marsh islands, zipping up Old Teakettle Creek. We sailed under the phone and electrical wires that led to Sapelo Island from Valona. Like church steeples on a skyline, I could make out the tips of outrigger booms on shrimp boats parked at the Valona docks. The vessels rested, but wouldn't be idle long.

We arrived at Rafe's spot in the Mud River with the tide still too high to start work. The water had ebbed enough for us to anchor by a wild oyster mound dubbed Old Faithful. With the footprint of a small house, Old Faithful had sustained Mike Townsend's shellfishing business for many years. But Mike had allowed Dom at the DNR to carve out a parcel for Rafe that included this mound in exchange for a remote spot where Mike thought he could best hide his farmed clams.

We all set foot on Old Faithful, a maiden landing for all involved. Stepping from clump to clump to avoid pits of mud, we perused the bounty. With time to kill, we helped ourselves.

Using a knife, Justin shucked a muddy oyster and handed it to Rafe.

"This is amazing," he said as he slurped it down. "It's crazy that this is mine."

Rafe grew up in Atlanta, in the wealthy community of Buckhead. With his father out of the picture, Rafe's mom worked multiple jobs to afford an apartment in that posh neighborhood and private school tuition for Rafe and his sister. Social status was important to his mother, Rafe told me later, which was why she worked so hard to keep them in a world they couldn't really afford.

"My sister loved it, but me, not so much," Rafe said.

Rafe met Ansley, heiress to a flooring and tile empire, in high school. They attended the University of Georgia together. Their postgrad life took them to Vermont and Northern California, with Ansley pursuing her creative passion—photography—as it blossomed into a career, and Rafe finding employment on farming operations, from organic produce to livestock pastures. Rafe schooled himself in running large-scale farming operations at a prestigious program at the University of California, Santa Cruz.

Qualified and experienced enough to run his own farm, Rafe and Ansley returned to Georgia. With the help of Ansley's father, they secured land in McIntosh County, and Rafe worked day and night to make Canewater viable.

Access to oysters were lagniappe, a product he wasn't sure would make him any money. It would embellish his business like a feather in his cap—authentic, but perhaps unnecessary. He had no idea how much work it would be. But Justin tried to encourage him.

"That's about as far as she's going to go," Justin said when we were all back in the boat, floating above a wide mudflat. At normal low tide, we would be beached on mud, unable to motor or even walk, the nearest navigable waterway glinting, mirage-like, to the west.

Rafe took off his pants to reveal a wetsuit bottom. In place of muck boots, he slipped on diver's boots. He looked ready for whitewater rapids, not bivalves.

Justin and Rafe jumped overboard into cool water that rose almost to their waists. I handed them PVC pipes, which they jammed into the muddy river bottom at two foot intervals.

Pipes secured and arranged in two rows two feet apart, they took the rebar I handed them and ran the lengths through holes cut in the PVC.

This construction took place below water. In moments when Justin or Rafe wasn't kicking up silt, I could see from the boat what they had built. This cobbled-together oyster-farming structure resembled the scaffolding in many muscadine orchards near my home in northern Georgia. Vertical posts connected by crossbeams, the fruit hanging in the interstitial spaces.

On the rebar, they fastened the square oyster bags with zip ties, each corner attached to rebar so that the width of the bags floated between the runs. At high tide, the river would keep it buoyant. At low tide, the bag would scrape the river bottom.

Each bag held 250–500 oysters. From this day on, it was up to Rafe to tend to them. Once a week at least, Justin advised, Rafe should come shake the bags. Every month or so, open the bags up, pull out dead oysters, and sort the living ones by size. Match slow growers together, quick

Justin Manley and Rafe Rivers in the Mud River.

plumpers together. For someone already running a bustling farm, this was a big ask, but Rafe promised to do it.

In the waning moments of their labor, the tide rose up to Justin's chest. The water soaked his cotton T-shirt. The setting sun turned the reflection of his sunglasses into fireballs.

The river, refilling ever faster, absorbed light into its murkiness like a black hole. The scene was both romantic and dangerous, and there was still work to do.

Justin had brought along a piece of ten-by-twenty-foot netting. Next time he came out to help Rafe, they would tie cages of oysters onto the net, which would keep the cages from sinking into the mud.

He and Rafe laid the net onto the river bottom, with plans to stab the remaining strands of rebar into its corners to secure it. Justin grasped a steel bar with two hands and thrust it into the mud like Poseidon wielding a trident. For Rafe, the same action did not come as easy. Justin helped. He took a length of rebar from Rafe, then cycled his thick deltoids through an arc to bayonet the rebar deep into mud bed. That was it for today. They were done and we could leave.

The ride back to Crum Creek felt faster on the incoming tide. For a moment, in the girth of Doboy Sound, we rode in the wake of a shrimp boat that looked ghostly in the gloaming.

Before that, we sped eagerly away from Mud River. Justin cautioned Rafe on how to keep his boat in the deep channel.

"Watch the crab pots," he said, pointing out the white buoys that marked the location of underwater crab traps. "Keep it between those."

"I have so much to learn," Rafe said. "All that information is just up there in your head, huh?"

"It got there the hard way," Justin said.

Justin spied a figure maybe two hundred yards ahead, a boat that had stopped abruptly with its pilot standing stiff.

"Slow down," Justin said. "I wonder what's up."

Rafe kept a distance between the two boats and called out to see whether everything was all right.

"That's Laddie," Rafe said to Justin, finally able to make out a face. Everybody knew Laddie, even part-time McIntosh residents. A river rat, by some accounts, he linked many people near Darien to fresh seafood on the cheap. How he did so was a matter of controversy.

"I think I hit something back there," Laddie yelled across the river. "Y'all watch out for it."

Rafe waved at him and moved us on.

"He's lying," Justin said. "There's no reason for him to be in this part of the river this late at night. He's up to no good. Why is he out here at dark? Ask yourself that. He's bootlegging."

Laddie, I learned later, had recently had a commercial lease stripped away by the DNR. Too many complaints from his colleagues about Laddie not being able to keep his paws off other watermen's oyster mounds. But the punishment hadn't stopped him from illegally harvesting oysters and selling them to vacationers in Darien.

"You're going to have to pay attention to those kinds of things. We've all got to look out for each other," Justin said.

Closer to Crum Creek, Rafe's poise while navigating the river cemented. This part of the marsh he knew pretty well. The rest of it would come later.

We unloaded the boat under the light of a lone bulb, said our good-byes. Rafe thanked Justin profusely, and Justin smiled. "It's my job," he said.

That wasn't true, not really. This wasn't Justin's job, it was his life. Like a kid on Christmas Eve, Justin hadn't slept the night before in anticipation of the day's labor at Rafe's. He loved the marsh, working out in the elements, and champed at a chance to do so.

But there was something more. Justin had undergone a personal evolution from grad student to Spatking to extension agent. All along, the rebirth of an oyster industry served as a unifying theme. His entrepreneurial efforts had ended without reward. This new life of service to oystermen, though, held promise.

"There are reams of research and approaches gathered by science, and it's my job to make it all accessible. There's nothing better than seeing someone like Rafe take to this and be excited about it," Justin said as we began the long drive back to Skidaway. "Some of the other guys, all they see is the work it takes. But they're clammers! That's about as hard as work comes. They don't realize that what they think is hard work is the same no matter where your oysters are. Up north, they've got all that ice. Here, it's mud and tides, but it's all hard."

He felt Rafe was the kind of guy who could put his head down and make this work. It didn't hurt that he had a millionaire businessman backing him.

"We need farmers, people who think like businessmen, and Rafe is a farmer," Justin said. "He's the future."

"Still," Justin continued. While optimistic about Rafe and aquaculture and the imminent hatchery commencement, he felt overwhelmed at times. This industry shift, played out in real time and not in his dreams, was becoming far more complicated than he had imagined. "It feels like this weight on my shoulders. It takes all I have. It's this big animal. It's heart and hands and head."

We kept quiet much of the way home, lulled into silence by fatigue and the numbing sheen of lights reflecting off interstate signs. Neither of us knew at the time how the project would play out, but after the excursion with Rafe, we both found it hard to feel pessimistic.

THE McINTOSHES

Earnest McIntosh Sr.

Justin could tell me very little about the McIntosh family. He appreciated the elder McIntosh's quiet demeanor and marveled at the beauty of their cerulean eye color, a rarity among African Americans. He figured the scattering of brilliant light across the iris stroma tied the family to the founding of Georgia—its history was embedded in their genes.

Check out their last name, he said: McIntosh, just like the county, just like Lachlan McIntosh, one of the state's founding fathers. Earnest, his family, and his community at Harris Neck weren't just the descendants of slaves, they were related by blood to the rich landowners who ran Georgia for generations, who sired children with the Africans they enslaved. Those blue eyes, those European eyes, Justin wagered, proved it to be true.

There was a problem: Earnest's eyes were green. The older McIntosh corrected me on the matter when I made the mistake of calling them blue in his presence. When he let me inspect them closely, there was no confusion—his eyes glinted a green that resembled the afternoon sun lighting up empty beer bottles. Nevertheless, the gravity that Justin intended to convey by noting Earnest's eye color outlasted the gaff. There were few families more genuinely coastal than the McIntoshes. This authenticity resulted from generations of injustice and pain.

The McIntoshes' connection to this place, I would learn, was stronger than any I had encountered when interviewing oystermen. With Mike Townsend, who shared the name of the largest town in northern McIntosh, I skirted the intricacy with which watermen were woven into their land's past. In Harris Neck, not one breeze blew through its woodlands without history coordinating its path. Slavery. Black land ownership. World wars. Environmental conservancy. Harris Neck's tale shared currents with many turbulent and violent American story lines.

Earnest, sixty-three, wasn't one to dig up ghosts or talk about the civil rights struggles and land disputes that have made Harris Neck national news over the decades, but he saw his work on the water as the continuation of a historical record.

By oystering, he kept threads of his ancestors' lives stitched into the present. His son would weave them through to the next century. I hoped to hear whether something relatively modern—oyster aquaculture—would play any part in that future.

Since Earnest and his son hadn't requested any extension aid, no opportunity presented itself for Justin to visit Harris Neck with me in tow. I went on my own. Earnest told me I could catch him coming off the water a little before noon. He instructed me to meet him at the Harris Neck dock.

From southbound Highway 17 not far from Riceboro, I turned left at the Smallest Church in America. Harris Neck waited at the end of the road. I passed homes of varying vintage and upkeep, the crumbling Eagle Neck grocery and tackle store, acres and acres of timberland, subdivisions with overgrown gated entrances, stretches of impenetrably thick forest punctuated by golden blasts of marsh spanned by earthen bridges, and a ranch home so crisply burnt only the bricks remained.

Then I arrived in Harris Neck, described by Works Progress Administration writers in the book *Drums and Shadows* as "a remote little settlement connected to the mainland by a causeway."

In the late 1930s, WPA writers came here to document the spiritual lives of former enslaved people. In Harris Neck, they encountered "a peaceful atmosphere about the entire island; life flows along in a smoothly gliding stream; the people seem satisfied for the most part with a simple, uneventful scheme of existence."

Earnest's house was easy to spot. A lawn sign announced that I was in the right place: Harris Neck Oysters, McIntosh and Sons, and Earnest's phone number block-printed on white plastic. Two of these signs stood upright on Earnest's freshly mowed lawn.

I stopped briefly to catalogue Earnest's house in my mind: a small abode of just over a thousand square feet, the block painted sea-foam green, the siding done up in a matching but much lighter aqua. Earnest was a carpenter, he told me later, and had built the house himself. A tall wooden fence hugged a roundabout driveway at the rear of the property. Behind it was an outdoor mechanic shop filled with boat parts, tools, and hulls. The McIntoshes fixed their own gear.

Near the fence gate, Earnest had built a roof and a platform that wrapped around a walk-in cooler. But no trucks or cars were in sight this morning, just before lunch on a weekday. The McIntoshes were still out on the water.

From Earnest's house, another causeway led to the easternmost stretch of Harris Neck. The road split upon reaching this last tract. One road took permanent residents to homesites to the south; some of those lots were described, without registration of the violent legacy of the word, as plantations. The other route veered north into a federal wildlife preserve that comprised the bulk of the marsh island's acres.

Sportfishermen's trucks followed the latter road toward a public dock that overlooked the Barbour River. A large parking lot located there offered plenty of room for extended cabs and boat trailers. Tourist cars could also be found, their occupants out stomping on the metal gangway so that the gold and green and deep dark marshland could reel out before them for family photographs, like a watercolor landscape they could actually smell. Just a few yards north of this beautiful vantage point, Earnest and the few other Harris Neck watermen kept their boats at a private mooring. Three, maybe four boats at most were tied up there. Nobody sat on it casting lures or tossing seine nets. It was a waterman's dock, and not for hobbyists.

With time to kill before Earnest returned from oyster harvesting, I took a spin through the Harris Neck National Wildlife Refuge. The road sent me through a mossy canopy of oaks, palmettos, and cypress trees—a beautiful but typical image in these parts. It was a sight witnessed time and time again: a basilica of leaves and limbs, greens and browns. But it never failed to awe.

The refuge's tranquility amplified reverberations of nature: calls echoed from bird beaks, the flutter of feathers leapt from tree limbs, signs warned of alligator threats, fish tails disturbed the calm of pond water. The living roof retracted to uncover what looked like a meadow. Upon inspection, another purpose was revealed. Tufts of blond weeds poked through cracks in long runways of cement. The flat trough between treelines extended toward the horizon, north and south. This was the remains of an airstrip, but no plane had flown here in more than fifty years.

Before it allowed plants and wildlife to thrive peacefully, protected from human encroachment, the refuge's land served as an army airbase during

World War II. Long before it housed the aircraft of the Third Fighter Command, the land belonged to the people of Harris Neck—the ancestors of enslaved Africans, the friends and family of Earnest McIntosh's forebears. They had not given up the land willingly.

In September 1865, two and a half years after Abraham Lincoln announced the Emancipation Proclamation, and months after Sherman issued Special Field Order No. 15, which gave hundreds of thousands of acres along the South Carolina and Georgia coasts to former enslaved peoples, a reading of the last will and testament of the plantation owner Mary Ann Harris deeded Harris Neck to her husband's former slaves.

As other plantation owners sold off lots to freed slaves, northern McIntosh became a thriving community. People lived in tandem with the wetlands. They found income and nourishment from a simple existence. Oystering and farming garnered modest wealth for Harris Neck's population. Canneries owned by the Oemlers and then the Maggionis kept hundreds more employed. By the water and their farms, they lived in relative peace for decades.

As World War II reached a violent pitch, the Department of War chose in 1942 to use eminent domain to assume ownership of nearly three thousand acres of Harris Neck land. The people of Harris Neck smelled corruption. Rumors had spread that warships had been sunk near Brunswick, and that German U-boats had been spied from the Harris Neck bluff. Those glimpses of bogeymen, residents contended, were used to lure army officials into acquiring black-owned Harris Neck. Why Harris Neck, they thought, and not the equally suitable, white-owned land just to the south?

Harris Neck elders told journalists in the intervening decades that McIntosh's white leadership didn't like blacks owning their own businesses, being self-sufficient, not needing whites. Economic power made Harris Neck's chiefs arrogant back then, their offspring told the *Florida Times-Union*, and the confiscation of Harris Neck was all part of a plot to show them their place. With many of its native sons, including Earnest's father, off fighting the Germans and Japanese, Harris Neck couldn't mount much of a defense against the takeover.

Airbase construction displaced dozens of residents, destroyed businesses, disassembled churches. One witness retold her story of the expulsion to the *New York Times* in 2010. In a race to beat demolition, people used mules or just their broad backs to tote furniture and keepsakes from their homes. Evelyn Greer and her mother rushed back to Harris Neck to retrieve a phonograph, believing they had time before their former dwelling would be set ablaze. They were too late. The last time they saw their home, sparks and flames were licking its frame.

After the war, the land was turned over to the McIntosh County government. Harris Neck residents believed they would be allowed to return once the airstrip was no longer in use. If that had been promised, they had been lied to. Many landowners were paid by the government, but residents today argue that Harris Neck's African American population received far less per acre than the few white people whose land was taken.

McIntosh officials mismanaged the land: buildings on the property were used as private clubs where white people gambled and partied. According to a 1982 article in *Southern Exposure* magazine, federal orders stated that the land could function only as an airstrip, nothing else. But county leaders parceled off the land to white citizens, who started farms, built churches and homes.

Before, Harris Neck had been a utopia. In the hands of white McIntosh, it became a playground. The displaced who remained could look on and shake their heads. After nearly two decades of county misuse, the federal government learned of county corruption and reacquired the land. The Fish and Wildlife Service took over, and old Harris Neck became the refuge it is today.

Long home from the war and now successful fishermen, a group of Harris Neck men, led by the sons of black watermen, launched a political campaign in the late 1970s to return Harris Neck to its original owners. Guided by the Reverend Edgar Timmons Jr., the son of Captain Sonny, the most famous black oysterman in Georgia before his death in 2009, they filed a lawsuit against the U.S government, seeking to reclaim the land. In an act of civil disobedience, Timmons and other Harris Neck residents, including at least one of Earnest's relatives, erected a tent city on the land. Only arrest would remove them from their protest camp.

A district court and an appellate court examined Timmons's lawsuit, and both found his argument without merit. Only a congressional act could return the land, the judges ruled. The tactic used by those still fighting today to get the land back is simple but arduous: convince Congress to pass a law giving federal land back to citizens. The campaign has continued and seems to be without end.

Although some of his neighbors continue to advocate for the land's restoration, Earnest has moved on.

"It really don't bother me, that was before my time," Earnest told me. "I'd like to see it come back, but I'm not going to sit here and die about it. I'm going to do my business and move on."

Some still believed the return of Harris Neck was possible. Environmentalists worried that the wildlife flourishing in the refuge would suffer under human hands. Those in favor of reparations knew that the people of Harris Neck once lived in harmony with the natural world and could do so again. The most recent plan lobbed by proponents included the development of only a tenth of the nearly three thousands acres to build an eco-friendly convention center, an organic farm, and other green attractions. It was hard to say whether a return to the animal-plant idyll promised by Harris Neck lobbyists would ever materialize. As with most things on the coast, change came slowly, if it came at all.

The U.S. government, which owned the dock where Earnest permanently tethered his boats, offered its use gratis to the native watermen of Harris Neck and their offspring.

"It's part of our pacifier. I don't know if I'm using the right word. It's to keep us on the humble side," Earnest once told me. "We have to have that to work. If we don't have that [dock], we don't have access to the water. If they take that from you, it's telling me that the government doesn't want you to survive."

Watermen were leaders in the community. It wasn't unfair for people like Earnest to believe the dock was a bone thrown by the government to keep them happy. If that was the case, Earnest understood the truth of the situation, accepted it, made nice with the refuge people, and moved on.

"Now I'm the oldest one down there, and I'm the spokesman for the community for the Harris Neck dock," Earnest said. "For me, you know, I don't have a problem with the wildlife people. They work with us. I tell anybody quick. Whatever comes up, they work with us."

Having left the refuge, I leaned on a metal railing that wrapped around a cement pad onto which trucks backed up. I looked down on the floating concrete dock, where work boats quavered in the tidal river. Auric cordgrass savannas unfolded between mud banks, wide and bright like shocks of honeysuckle. Palmetto hammocks disturbed the expanse like peacock feathers, as did the incoming sputter of an outboard motor.

Three bodies in a deep-hulled boat laden with hundreds of pounds of oysters approached from the south. The vessel meandered toward the dock without a rush. I felt an urge, as a writer, to equate the speed with which the boat plodded toward its anchorage to its pilot's possession of a relaxed frame of mind, that he was in no rush.

Earnest, I would learn, chose to move slowly as a matter of calculation, not leisure. He planned everything—each decision and physical action resulted from great internal debate. But slowness on this day, it turned out, resulted from a mechanical accident. The sluggish pace didn't indicate a pleasure cruise. Something was broken.

Earnest Sr., Earnest Jr., and a nephew arrived at the dock and announced that a cracked bushing had delayed their return to Harris Neck by retarding their speed. They would fix it later, since other tasks were more pressing. Working as a team, they unloaded bushel after bushel of oysters, laying the bags in a long straight line on the dock. They peeled off layers of rain gear. Both McIntoshes were dressed similarly in collared shirts underneath monochrome sweatshirts. Senior's was green, Junior's blue.

Earnest Sr. unfurled a hose, turned on a spigot, and began spraying down the oysters. Mud dripped from shell crevices inside the bushels and washed over the side of the dock. With one ablution complete, father handed the hose to son, who sprayed down the boat until the last brown glops disappeared.

The cousin flipped over the bags, and once Junior finished cleaning the boat, he continued washing down the oysters as Earnest Sr. watched. (The oysters would get another wash back at the house.) Earnest Jr. then ran up the dock and backed up his truck onto the cement pad above. He switched on a large electric winch and lowered a pallet, bound to the lever arm by chains, onto the dock. All three men heaved the bags onto the pallet, forming a pyramid of bushels.

"That's about seven hundred pounds of oysters right there," Earnest Jr. said. His facial hair was cut into a Fu Manchu, and he wore a plain ball cap. "You had an oyster from around here?" he asked me. "There's nothing like it.

"They're clean and fresh," his father added.

I had seen references to Harris Neck oysters in newspapers dating back to the nineteenth century. Even then, the superiority of the product read like holy writ. The McIntosh clan believed that what was written then was still accurate today. I asked whether he thought there was a noticeable difference between the oysters around here and those in Mud River or Liberty County.

"One hundred percent," Earnest, Sr. said.

I chatted with Earnest Jr. as he operated the winch, pulling up the load of oysters and setting it into his truck bed, about whether he saw a future in oystering after his father retired, whether oyster farming held any promise for him. Below us, Earnest and his nephew collected gear and wet clothes and trotted them up the gangway from the dock. Earnest Jr.'s response was short but thoughtful. A toughness girded his voice, but kindness was the source of its strength. He looked at me with eyes as striking as his father's. His were emboldened with a youthful passion.

Earnest Jr., only thirty years old, had four young children at home. His wife had a college degree, but that didn't mean much in McIntosh County, he said. He held his family's financial security in his hands, but he was hopeful, not scared.

"It's the future, you know?" He said. "We'll have these pretty oysters, and we'll get them into restaurants. That's where the money is. There'll be a future here for my son."

His father had fewer revealing statements for me, since they were in a hurry to clean these oysters and deliver them to Charlie Phillips.

"You probably got what you needed," Earnest Sr. said to me as he closed the door to his truck and put the oyster-filled vehicle into reverse.

"No, but enough for now?" I said, with a look that asked whether I could interview him again. He nodded and drove away.

SPAWN AND SPUTTER

Justin Manley watching oysters spawn.

Justin raised a mattock in a striking arc. The sharpened tip sunk into sand grains and sediment like a warm spoon into ice cream. With a thwack, the tool met its intended target—a six-inch-wide root, part of an underground system that burrowed like a neural web between the shellfish lab's foundation and the riverbank.

Justin lumbered through this work as if it were part of a waiting game. He needed the hatchery to be online by early summer, but important system parts were on back order. It was March, not yet hot, and not too late in the process. Justin didn't mind waiting, since it wasn't a fallow period. A trench had to be dug. He exhumed roots and dirt to make room for plumbing. Before the downweller tubs arrived, Justin had to install a pump that would draw brackish water from the river through pipes buried in the trench. In the lab, the water would condition oysters, flow over and around them, mimic the wild river. For now, that room sat empty. If Justin was to be ready to spawn the first batch of hatchery oysters on time, a few things had to go right, mostly in the realm of shipping and delivery. He was confident all the dominoes would fall as planned; he had prepared as best as he could. For now, mattock whacks and shoveling offered a break from the details. The brute labor relaxed him, gave his muscles purpose. And they would get plenty of purpose: when I came upon him in the morning, he had progressed ten feet from the back wall of the lab—ninety more feet to reach the Skidaway.

I found it hard to let Justin work alone, to scribble notes as another person hammered along in front of me. I offered to assume the yoke and give him a break. He did so happily. But although I was once a house framer, I found that my muscles brandished less power than they used to. I couldn't apply the same force as Justin. But I tried. My blows nicked root skin, but they refused to sever during my shift. Justin let me bumble on until I realized I was wasting our time. I handed him back the tool.

"I warmed them up for you," I said. He laughed, then proceeded to cut through roots like a wheat thresher at fall harvest.

A few months had passed since Justin and I spent extended time together. I used the winter to pay visits to watermen to the south while Justin readied the shellfish lab for its new role as a hatchery. The project focused him, but I could tell the sheen had begun to fade. Tom Bliss hired a new extension agent, a twenty-something biologist named Rob Hein, a Georgia boy who had just returned home after working in the Caribbean among sea turtles. His arrival meant Justin could concentrate on the hatchery while Rob performed the fieldwork. This was the plan all along, but the shift disconnected Justin from the watermen. Through the previous summer and winter, he had strengthened relationships with "the guys," and now he had to back off. Luckily, the quotidian tasks of starting the hatchery occupied his day and mind, but it was obvious that the enthusiasm I had witnessed on the water with Rafe had faded. The romance of water work evaporated on dry land. He had good days and bad ones; some days full of resolve, doubt on others. Some days he would give anything to trade the moil of the lab to sit a tide. That was certain. Other days, he found serenity in to-do lists and spreadsheets. He dreamed of tides, but they stayed at ebb. Tidbits about watermen's trials and triumphs reached him, but it all came secondhand.

Justin asked whether I had any news. What had the guys been saying? Were they optimistic? I relayed the positive and the negative, that Earnest Jr. seemed excited, and that pretty much everybody felt daunted by the constant gardening that aquaculture required. Justin had predicted their skepticism but didn't know what else he could say to ease their minds.

Perhaps the repetition of ditch digging wore him out, but I sensed frustration: "The guys don't realize that what they think is hard work is the same no matter where your oysters are," Justin said. "Up north, they've got all that ice to break through. [Here,] they've got their oysters in racks, and they use a winch to haul them up, and they pressure-wash the mud off. There's nothing to complain about, but everybody does, especially because they can't see the payoff."

I told Justin about my January visit to Dom at the DNR's Coastal Regional Headquarters in Brunswick. I had found the conversation informative and noted that Dom's approach to his job was more nuanced and philosophical than I had expected from a state bureaucrat. My interactions with him and other watermen during the last GSGA meeting offered a brief glimpse at the harmony Dom was attempting to arrange between the needs of his department and its primary stakeholders, the oystermen. As Dom answered the long list of questions I had prepared for him, I came to appreciate the tricky political nature of his appointment.

The DNR building stood in the shadow of the massive, marsh-spanning Sidney Lanier Bridge and reflected the winter sun with dozens of windows. Seventeen steps led into the building, a blend of Greek Revival columns, common in the South, and glass-encased modernity. Inside, announcements about fishing regulations and license guidelines, along with informational posters, plastered the reception area walls. I waited there for Dom to greet me, reading about catch limits for speckled trout.

Dom's office wasn't in the main building. He walked me through labyrinthine hallways, down stairwells, and out the back door. Abutting a creek, a structure that looked as much like a marine machine shop as it did a governmental agency sat across a stretch of cement. We crossed between buildings. We walked up to the second floor and reached Dom's windowless, musty, wood-paneled room of file cabinets, binders, and maps.

Dom praised Justin's mix of education and drive, and described himself as an advocate as much as a regulator. "Oystering has all the right things behind it, it's why I come to work," he said. "I truly believe in the validity of the industry." We talked of theft, oystering gear, and Crown grants. He told me of all the emails he had received from giant aquaculture operations from Alaska and the Northeast looking for leases on the Georgia coast. He respectfully, and digitally, closed the door in their faces. He explained his professional preoccupation with public health, so I asked him a question often posed to me: why were laws stricter for oysters than for clams? The main difference, he said, is that oysters are eaten raw. Cooking erases many shellfish dangers to humans. But as more restaurants began offering raw clams, quahogs would come under similar scrutiny. He was part of a team, he told me, tasked with revisiting public health protocols surrounding clams. Change was coming.

But what of the future? In the transition to aquaculture, which, he believed, was squarely underway, he wasn't looking for leaseholders with knowledge of wild harvest. Poor coastal people had been great stewards of the resource in the past, he said, but the new reality of farming required big plans and financial backing. This didn't preclude traditional fishermen, he said. "We're not removing the old ways," he said. "People are always going to want a cluster oyster."

As retiring watermen gave up leases, Dom planned to split them into smaller parcels. Carving fifty acres out of Mike Townsend's lease for Rafe had been an experiment, and it seemed to be working. "With mariculture, the domain doesn't need to be as large," Dom said.

All this was new to him, he told me. He was learning alongside everyone else. He had no crystal ball, he joked, so he had decided to take it slow, even though he would face criticism from watermen for not making decisions on things like oyster-growing gear as quickly as they would like. "I don't want to outgrow my ability to manage the resource," he said.

I reported all this back to Justin. He agreed with Dom's plan to not take away acreage of current growers and to cut up leases only as they came open. Let the wild harvesters alone, let them go their own way, Justin said.

As he shoveled and cut roots, we discussed an organic chicken farmer, a small-scale producer raising his birds on pasture in Reidsville, about forty-five minutes from the coast, who had expressed interest in oyster farming. The farmer had already contacted Justin and Dom; his application neared the top of a waiting list. He was another strong candidate for oyster farming: environmentally aware, known to city chefs, a marketing whiz. Between Rafe and the chicken farmer, Justin felt his forecast about the future of the industry was being proved true. Clear the way, and a new breed would move in. He hoped, of course, the incoming class would share Rafe or Earnest Jr.'s agrarian tendencies, rather than resembling one of the faceless mariculture conglomerates that pestered Dom. Just as Justin, Tom, and Dom urged the creation of a line of endemic oysters, they also hoped an oysterman of similar pedigree would raise them.

A native oyster farmer would develop at his own pace, but Justin already had bivalves eager to spawn. He wanted to introduce me to his brood stock. We left the mattock and the trench and the daylight dappled by oblong leaves and walked toward the dock—across a boardwalk, underneath clouds dabbed like milk splotches onto a cerulean pool, and down a gangway to Justin's floating upweller. Hard plastic canisters called tumblers hung from the dock by ropes. Justin bent down to dip his forearm into the river. He hauled one of the canisters onto the dock. Sea grass dangled from the black frame. He unlatched a side and shook the bag. Three dozen oysters tumbled onto the dock. These were his wild stock, collected on spat sticks not far off the dock in the Skidaway River. These were his prized performers, the plumpest in the bunch. He held up a fat one for me to inspect, an oyster like a dog's paw, slate colored and black streaked. He admired its cupped teardrop shape; the meat inside would be stout and round, he promised. He described the specimen as oviate. He traced a finger along the shell's translucent tip, which glowed white hot in the sun. He dared not touch it, knowing it would cut him. The sharp and brittle end told Justin the oyster was maturing healthily. Big and getting bigger, it displayed all the characteristics he had hoped to produce. To Justin, it was perfect.

"We'll get a female and a male of these and hope they spawn a new generation," Justin said. What he didn't know was which sex he held in his palm. Several environmental factors determined sex. Over its lifespan, an oyster switched sexes, Justin said. If an oyster sensed the presence of too many males, it would become female and produce eggs at spawning. But without shucking the oyster during spawning season and slicing a razor into the animal's gonad, effectively killing it, he had no way of knowing the sex. That he lacked this information didn't worry Justin. Sex assessment was yet another step in a process with which he was becoming increasingly comfortable.

Justin's manner lacked the cocky bluster present in our previous interactions. His mood was calm; he spoke less than usual. I asked about his family. They were fine. The wife, kids, and he were all in a routine. Each day, Amelia left early for work, he took the kids to school and then headed to the gym to lift weights. He spent the rest of the day at the lab, tinkering until suppertime.

After fall's burst of wild spat delivery and farm construction, the pace had slowed. Yes, Justin's role as hatchery manager required ditch digging but also long sessions in desk chairs, eyes reddened by a computer screen. The excitement of the water was gone, and he would have to get used to it.

In the period between Test America and Justin becoming an extension agent, the Manleys had left Whitemarsh Island and moved into a house of their own not far from Justin's new office. John Pelli, a waterman and one of Justin's friends, lived just two doors away.

Under Amelia's name, they bought the house in 2011. They plugged away at renovations over three years, with Justin, his father, and family helpers running wiring and sanding drywall. Shielded by a mossy canopy, the lot backed up to creeks that wound through marsh grass to reach the Vernon River. The wood exterior and shingle roof look unblemished, as if the siding had been patched but not replaced. The interior glowed white like a Florida vacation home; soft light lit eggshell walls with brilliance.

Renovations weren't complete, Justin told me, but I found it hard to locate the unscrewed faceplates and untrimmed doors he promised were on his to-do list. Had there not been shoes arranged on the doormat, and family pictures hanging from the wall, I might have mistaken the house for a vacation rental. It was that tidy. But any sterility the home's radiance implied was made comfortable and homey by the buzz of the Manley family.

Their elementary-school-age children, Bella and Mercer, flew through the house, invigorated by a burst of after-school energy. They bounced from den floor to couches, into bedrooms then out to the backyard, which stretched out with fifty yards of Bermuda grass before dipping into the marsh.

Big blue eyes like tractor beams, short platinum curls bobbing, Amelia squealed a hello and pointed at wine bottles and empty glasses. Red and white, she had them both, whatever I preferred.

"I love real hoppy craft beer, but I can't drink more than one or two a night," Justin said, uncorking a bottle of pinot noir. "I can put back some wine."

As we talked, Amelia spooned mounds of rice onto plates and ladled chicken curry on top. She called the kids to dinner.

Bella and Mercer squirmed in their chairs, picked at bits of chicken, and pushed red pepper slices over errant grains. Amelia asked me whether I had visited Pin Point yet. There was an old oyster cannery just a mile down the road called A. S. Varn and Son, located in Pin Point, an early settlement of land-owning Gullah Geechee people. It was also the birthplace of Supreme Court justice Clarence Thomas. Just two years earlier, a museum had opened, the dilapidated shucking and packing houses rehabbed as exhibits. She said I would learn plenty about oystering history there if I went, and she was right. When I visited weeks later, I was finally able to physically connect with Georgia's oystering past. I stood over the bateaux that Gullah oystermen paddled out into the marsh in frigid winter, warmed only by charcoal sizzling in a smudge pot. I stood in the same spot where Pin Point women had shucked oyster after oyster, fingertips frozen by winter air and packing ice. After an 1890 hurricane forced the Gullah Geechee to flee Ossabaw Island, they settled in Pin Point, becoming some of the earliest and longest-lasting Gullah Geechee landowners in the Lowcountry. Mr. Archie, as A. S. Varn, a white man, was known to the people of Pin Point, paid bills, advocated for youth, and supported Gullah baseball teams. His grandson, Algernon Varn III, who still lives on the property, recalled how the cannery, on off days, served as a social center for Pin Point; oystermen and canning women cooked, danced, and celebrated life there. The cannery closed for the same reasons all the others did: development like the Diamond Causeway, which connected Skidaway Island to mainland, locals claimed, negatively affected marsh life; younger generations left for work in the city. Pin Point trudged along, its houses studded by recurring moss bloom, as the world modernized around it. Just one mile away, a new breed of oysterman—a man born miles away from Pin Point—had established his homestead.

At the risk of starting an argument between the couple, I asked whether she missed Justin working on the water. Yes, of course, she said. It made him so happy. Selfishly, she said with a smile, she missed eating what he grew. "It was so nice having all that shellfish," she said.

"I'm a guy," Justin said. "I'm hardheaded, but I have that manly pride. There was no way I was going to let you carry this family all alone."

She rolled her eyes at Justin's dramatic turn. On the point of providing, he didn't budge.

Justin prodded the kids through their bedtime rituals while Amelia and I opened another bottle of wine. She leaned in to ask a question. "You like oysters. Have you ever thought about doing it yourself?" she asked.

"It's tempting," I said. "I'd love to have to be on a boat everyday."

"I want him to have his own farm again. He really needs it," Amelia said. "Maybe you could do it with him, be his business partner." Sure, it was his project, the Spatking Oyster Company, but she took some ownership. She was proud of Justin's entrepreneurial efforts. She loved the idea of the Manleys owning their own business—it had always been a dream of her own. "If I had the skills, I'd do it, too," she said.

Justin returned after tucking both kids into Mercer's bed so that I could sleep in Bella's room. Modern jazz played through the satellite TV. A marsh sightline was lost in the recess of dusk.

"How much money would it take to get an oyster farm up and running, at a minimum?" I asked Justin.

"A hundred fifty thousand dollars would be a good start," he said. I told him what Amelia had asked me. He shrugged at first, then said, "Sure." I could tell he was laying out a plan in his head. An index of to-dos escaped his mouth: a flurry of what we would have to do first, what stars would have to align. "They want young people doing this," he said. "You might be a good candidate for a lease."

He quieted himself but continued to mull over the idea. "Maybe down the road," he said.

As more wine legged down the sides of our goblets, the conversation veered toward lighter topics: hangovers, Montreal, the best movies they had seen recently, our children.

I woke early, before dawn, to drive home to Athens. Justin was up making eggs for the kids and coffee for me. I declined breakfast, and Justin and I exchanged good-byes with bleary eyes.

Back home, I surrendered myself to the fantasy of working with Justin. How reasonable would a career change be? Sunburnt, salt dried. I romanticized its potential. It was a temptation that I figured many writers had—to dip toes into the worlds we witness and research. I even asked my spouse whether she would ever consider moving to the coast. In the end, I recognized it smartly for the folly it was, however enticing the prospect might have been. I envied the labor of watermen, but it was not my path.

Yet the conversation did introduce the prospect that Justin might one day return to the water. I wasn't the only one pushing for that reunion; Amelia dreamed of it, too. Perhaps we hadn't seen the last of the Spatking.

EARNEST IN THE CITY

Earnest McIntosh Jr.

Justin eagerly informed me that Harris Neck oysters were now available in downtown Savannah. A new restaurant had opened, the Grey, and it stocked Earnest's oysters, serving them on the half shell. Not since the Spatking had this happened. But there was a difference: the Grey wasn't stocking farmed oysters, just wild ones. Earnest had impressed the owners with the cleanliness and smoothness of his wild singles. The partnership appeared to be permanent, Justin said, not seasonal, and definitely not just for show.

When I went to the Grey to investigate and meet its chef, it all made sense. Housed in a renovated Greyhound station, the Grey had quickly become a jewel in Savannah's crown. Much of the previous interior decoration remained—most notably, the metallic edging around wood paneling—and the designers hadn't sought to erase its ghosts. The worn floor around the ticket counter wasn't buffed to a modern sheen, and signs pointing to Jim Crow–era segregation still hung.

Behind the former ticketing window, cooks draped in white robes worked in a fury. The staff had placed their oyster selections near the separating glass for diners to gaze upon; the names of the oysters were written in neon marker on the translucent barrier. Harris Neck, in an orange cursive, $2.25 each.

One person could explain the appearance of a rural, local oyster in as fine a restaurant as the Grey: Mashama Bailey. Born and raised in New York City, Mashama had moved south a year before the Grey's grand opening. She had ties to the area, having spent a year in Savannah as a child, close to the hometowns of her mother and father. Now forty-one, Mashama had come to Savannah at the request of John O. Morisano, who wanted her to run his new restaurant, which promised to be a showcase not only for the chef who helmed it, but also for the city. Savannah was destined to return to the cultural spotlight, and the Grey would be its prized table.

Mashama carried plenty of baggage as she accepted the position and migrated south. She was a female executive chef in a field dominated by men; she was an African American executive chef in a field dominated by whites. She did not fill shoes, she blazed trails.

Her goal wasn't to be a figurehead for race and food and the South. She merely wanted to recognize, rejuvenate, and reinvent the cuisine of one of America's oldest port cities. No small task.

When she opened her notebooks, tied an apron around her waist, and got down to the business of writing a menu, contracting with food purveyors, and opening a restaurant, her problems weren't theoretical. They were very local.

Where in the hell was she going to find fresh seafood?

The first contacts she made in the Savannah food world delivered her only frozen eels and clam bellies, nothing that had been swimming recently. She found a good butcher and an organic farmer, but local seafood evaded her.

"Coming from a place like New York City, there's a person for everything, and I was looking for that down here," she told me while we talked at the Grey bar as service wound down on a weeknight.

As she was beginning to learn, in the Savannah food community, it was about whom you knew. For fresh seafood, the purveyors she needed to know didn't live in Savannah. Through a series of phone calls and meetings, which began with getting to know Cynthia Hayes of the Southeastern African American Farmers' Organic Network, Mashama began a journey that would lead her to Harris Neck. She traveled to Brunswick and St. Simons Island. She found shrimpers and fishermen, but with oysters, she struck out. Around Darien, she met a man named Wilson Moran, a pillar in the McIntosh County African American community.

She sat down in Wilson's living room, which happened to be only a few yards from Earnest's front door. As Cynthia Hayes, her guide thus far, and Wilson's wife chattered on, Wilson eyed Mashama.

"Hold up, you two," Wilson said to the gossiping women. He looked at Mashama: "Who are you and what are you doing here?"

Mashama, assuredly, was a stout woman. A corona of curly hair encircled a kind face not often found in bustling professional kitchens. Her thick-framed glasses conveyed a false sense of meekness. But she was not a woman easily rattled. And Wilson did not rattle her.

"I felt relieved because he didn't beat around the bush, he wanted to know what I wanted," Mashama said, her voice lilting in tones both southern and Bronx.

"My name is Mashama Bailey, and I'm from New York City, and I'm coming down to Savannah to open up a restaurant," she plainly told Wilson.

Okay, he said. What do you need?

Crabs. Oysters.

Well, Wilson told her, let's start with Earnest.

A phone call, a wait, and a returned phone call later, Mashama met Earnest. She did so not by walking a few doors down to his place, but by hopping in Wilson's car and driving the short distance. She was told that Earnest was peculiar, didn't like people just walking up to the house. If she wanted to meet him, they would have to drive over.

She met Earnest McIntosh Sr., likely dressed in slim slacks and a conservative shirt, in the comfort of his home.

"How old are you?" he asked. "Are you a student at SCAD [Savannah College of Art and Design]?"

"I hear that Harris Neck oysters are the best oysters," she told him. She didn't know what caused it—her spirit, her aura, something else—but Earnest responded.

She asked Earnest to show her his operation. I want to sell your oysters. How can I do that?

Earnest said okay.

From that meeting, a partnership began. But like any relationship, it took time for the allies to feel each other out. The first bushels Earnest delivered to the Grey weren't exactly what Mashama was looking for.

"It was these big clusters, and we would shuck one, but it was this mess, and we would lose some and couldn't put it on the plate," Mashama said.

After a time, Earnest learned what Mashama needed. She needed singles. With a little effort, he gave them to her.

"Now, it's a lot more focused. He knows what I want, and he can give it to me," she said.

Earnest's oysters would become part of Mashama's mystique. She would go on to embed an integral element of McIntosh County into her rise as a chef, helping Harris Neck oysters find mention in articles printed about her in the *Washington Post* and the *New York Times*.

I shared Justin's excitement in seeing Georgia oysters vaulted back into the fine-restaurant realm. Perhaps Mashama, finally, had performed the needed culinary work in seeking out vernacular ingredients and dragging them into her kitchen. Just a few years before, Justin had forced the connection from the other direction, transporting the marsh to the city. Mashama's approach seemed easier, at least for watermen like Earnest. Together, oysterman and chef, the linkage might last.

Additionally, Mashama's increasing acceptance by local and national media was the promotional ally the Spatking never found. As her profile swelled, so did Harris Neck oysters'. Maybe Justin had moved himself out of the way just in time to let a demand-side trend correct the oyster dearth.

Without aquaculture, without UGA, without Justin, Georgia oysters were again getting their due, thanks to Mashama and Earnest. Encased in chip ice in a premier restaurant, served to well-dressed eaters in a gorgeous dining room, it sure seemed that way.

Did Justin have nothing to do with this? Could the industry rebound without him, without marine extension?

Perhaps I was being too generous about the impact the Mashama-Earnest duo could have. Together they had trained a spotlight on Georgia's oysters, but a restoration of the industry still hid in the shadows. Theirs was an important step, but not a victory lap. It would take more than one restaurant to return Georgia oysters to prominence. After all, shrimp and clams still ruled. But at the Grey, oysters scored a choice reservation, one denied them for so long.

I returned to Harris Neck that winter on a day marred by rain and thunder. The poor weather kept Earnest and his crew off the water. Without the distraction of harvests, maybe he would answer my questions.

Earnest Sr. and Earnest Jr. had spent much of the morning scooting between parts distributors. The plan for the rest of the day was to make seaworthy a boat that Earnest had been keeping behind his house. When I pulled into the driveway, parking my car at the base of the small packinghouse, I could hear the screech of a grinding blade—metal on metal—and the whir of a drill-powered hole saw as it trepanned fiberglass into a scree of shavings.

The interstitial mist between rain gusts forced Earnest Sr. and I onto the covered deck of his packinghouse, the door of the cooler to our rear.

Earnest Jr., a cousin, and his uncle fretted over the boat, warmed by a fire burning in an oil drum.

How did he get his oysters pretty and clean enough to call them singles? I asked Earnest Sr.

"You get on a bank and you start breaking them loose, they lie down, and they become singles," he said. Let them be for a few months, then come back and get them. "You break down the small ones to make one look decent."

I told him about references to Harris Neck oysters in newspapers printed more than a hundred years ago. He nodded as if he had been in the archives himself.

"Oh, I know," he said in a voice sweet and paced and purposeful, like wood smoke infusing a whole hog with a hickory lacquer. "Harris Neck's always been quality."

Earnest inherited the trade and a commitment to quality from his father, a serially successful entrepreneur.

One of fourteen children—seven boys and seven girls—Earnest joined four of his brothers in the booming family business when he climbed aboard a crab boat in 1978, not as a boy traipsing after his father, but as a professional waterman.

"I'm the only one still at it," Earnest said.

Back then, the McIntoshes focused mostly on crabs. They harvested oysters too, but soon after Earnest started in the business, thousands of oysters died off from a protozoan attack that affected most of the Georgia coast. The wipeout caused Earnest to quit oysters. He didn't return for twenty-five years.

"It took that long for them to come back," he said. Luckily, Earnest's oystering homecoming coincided with a decline in crab populations. Some cited a multiyear drought in the early 2000s as the culprit. Faced with a crab shortage, competition among watermen became fierce. A fellow crabber pulled a gun on Earnest one day in the marsh, Earnest told the *Savannah Morning News* in 2003. Back then, Earnest expressed dismay for the future.

"The crabs are not there, and they're not coming back," he told a reporter.

The crabs would come back, but in nowhere near the numbers that watermen once hauled over the gunnels of their skiffs. No matter what had caused the decline, the slump prompted Earnest to seek other catches to make up for lost harvests. He looked to oysters.

"We needed something to do when the crabs were out," he said.

Years before the crash, crabs had kept the Harris Neck economy and the McIntosh family afloat. In 1980, the crab business was so lucrative that Christopher McIntosh, Earnest's father, broke ground on the McIntosh Crab Plant, consisting of four cinderblock buildings situated behind the home where he and his wife, Marie, raised their brood.

Earnest pointed me toward the plant's remnants: "Down there on Blue Crab Lane, that's our old home spot." I found one building still standing, the block painted red and the roof needing new shingles. Someone unrelated now lived inside the homestead.

Christopher and his sons self-funded construction of early phases of the crab plant. But to complete the project, they turned to Fort Valley State University, the Small Business Administration, a bank, and the McIntosh County government to raise $105,000.

When a writer for *Black Enterprise* magazine toured the facility in 1982, the McIntosh Crab Plant numbered fifty employees and owned five crab boats. The plant bustled with packers, baggers, watermen delivering crabs, and office staff filling orders. According to Christopher, there was room for expansion.

"Our biggest problem is a labor shortage," he told the writer.

For Christopher and his family, the crab plant represented a lifetime of effort. Earnest's father had started out as a carpenter to make ends meet, working the water when he had time, for extra cash and sustenance. Then he moved into migrant farmwork. As his family grew, the McIntosh clan trekked up and down the Atlantic seaboard, following harvests. Tomatoes, beans, potatoes, whatever the season was, Earnest Sr. said.

Other Harris Neck families followed the McIntoshes, and Christopher served as a labor contractor, supplying able bodies for the fields. Money saved from those agrarian endeavors speeded his return to the water, full-time.

Harris Neck's spirit, Earnest Sr. told me, always came from the water. It fed people, gave them jobs, made plenty of families happy.

"Around here, that's all there's ever been, seafood," he said. "There used to be ten or fifteen boats down there at the dock."

For his father, there was no place else.

"I've been a crabber and a river man all my life," Christopher told *Black Enterprise*. "A man needs to stick with the thing that makes him happiest, and my boys feel the same way."

Despite the physicality of the work, the nature of it appealed to Earnest. Put in a day's work, don't waste your money, save it, and financial freedom is yours. A rather honest way to make a life.

"I could go and work hard, and if I worked hard, I could scuttle together money to buy what I wanted," he told me. "I like it. It's a peace of mind. You leave, and you've got problems at home, you have things that aren't right, you get out on that boat, and you can talk, meditate with the good Lord. My wife used to get upset with me. She said, 'When you get on that boat you get a peace of mind.' It's the truth."

Earnest Jr. had tried to get away, but had come to the same conclusion as his father. Although he had chosen to raise his family in Darien, which he called "the city," nothing beat a life working on the water.

The younger Earnest was a standout athlete in two sports in high school. His prowess in football attracted the attention of a small college, and Junior left Harris Neck on a scholarship.

He hated being away, disliked college life. Not long after his departure, he came back. He worked for a time for Sea Tow, dragging boats with busted motors back to docks. It became his last effort at avoiding what he tried not to admit: working alongside his father, in the same profession as his grandfather, just like all the other men who had raised his friends and cousins, made pretty good sense.

"You know, he never really parted with it," Earnest Sr. said of his son. "It was still there, in him. I raised my kids like that, from the water. He's pretty smart about it."

Despite the clarity and profit Earnest Sr. and Jr. found on the water, Harris Neck had changed plenty since 1978.

He yelled at his son: "'Earnest, not but five kids who catch the bus of Harris Neck now?"

"Three or four."

"A lot of people used to be here," Earnest Sr. said. "The school bus used to come down here, park right over by our old home spot [he lived there until the mid-1990s]. School bus used to leave out of here, and it would be buck loaded, loaded all the way back with kids all from Harris Neck."

Wasn't much to keep people here, he said.

In McIntosh County, jobs off the water were scarce. If a young woman or man wanted a career not sweating or freezing or otherwise stretching their bodies to the limit, options were limited.

"There aren't enough jobs in McIntosh County," Earnest said. "It ain't Harris Neck's fault. It's the county. Not everybody wants to work on the water. Ones that [are] blessed to have the skill to work on the water, they're fine. But the ones who want plant work, they've got to go to other counties to get jobs."

Just as his father complained of a labor shortage in 1982, Earnest found young people in his corner of McIntosh to be uninterested in fishing for a living.

"Ain't too many young people who want to work hard," he said. There was his son, a brother, a nephew or two—that was it. It was enough labor to keep McIntosh and Sons Seafood going. For an enterprising person, a living could be made, Earnest Sr. contended.

Did he have advice for oystermen, for anyone who wanted a job from him, or wanted to get into the business?

"Pick good-quality stuff. Whatever you catch or pick, make sure you put a good product out. Be honest with it. Don't get over on nobody. If you pack a good product, they come back. If they keep that in mind, that'll take them somewhere.

"Here's what I tell my people: 'You've got to look at what you want to eat.' You get on the bank and you pick it. If not, leave it. That's what I tell the guys who work with me: 'Pick what you want to eat.' If you want to eat something bad, that tells me what you're about. Most of the time, if you're having an oyster roast, you try to pick the best thing on the table. Pick how you want them to look. That helps."

Earnest hoped to see Harris Neck thrive again, and seafood and good water jobs could bring that about if people committed to the place.

"The seafood industry is much higher than it used to be. It makes me want to stay into it. I can see a future in it," he said.

Would he ever retire?

"I'm here until I can't go again. I want to be in it for a long time, and I want to pass it on to him."

The clap of a bolt ricocheting across a five-gallon bucket drew my attention to the boat. I saw Earnest Jr. leading his cousin and uncle in the repairs. He wasn't just heir to the marsh off Harris Neck, he was his father's partner.

How the marsh and its people entranced me like mystics, how I respected Earnest's leadership and professionalism, how I belly-laughed when in Joe Maley's presence, made me worry that maybe something could be lost if wild harvest was abandoned in the quest to establish aquaculture. But in my time with Earnest, I realized that if anything could harm the character of Harris Neck, the hardest blows had already been landed.

He and his were fine. Maybe they would farm, maybe not. It didn't necessarily matter. The McIntoshes would find a way forward, be assured of that. A foundation had been laid by their ancestors, and it supported them ongoing.

I left Earnest's and took another drive through the wildlife refuge and ended up parking at the watermen's dock. I walked over to Gould's Cemetery, a sacred-feeling graveyard with headstones dating back to the 1800s.

McIntosh. Thorpe. Moran. The great names of Harris Neck engraved on small stones inscribed with stars. Edgar "Sonny" Timmons, the great oysterman of McIntosh, his grave site freshly flowered and cleaned. Christopher McIntosh Sr., Pvt., U.S. Army, World War II. Born Christmas Day 1920. Died May 22, 2008.

Shrouded by an umbrella of thousands of oak leaves, the graves overlooked the river. Decades earlier, there had been a church nearby and a landing. Like the deceased, it was all memory now.

I walked to the dock to look out over the Barbour River. I peered down at Earnest's boats as rain pelted me.

Gray layers enveloped acres of salt marsh that spread out between the mainland and St. Catherines Island, which looked as faint as a pen stroke across a monochromatic landscape. History enrobed this place in a heavy fog, and there was a sense that intangible truths could be clutched if you leaned far enough over the railing. Somewhere out in that silvery time-stopping sweep, Earnest kept his peace. As I craned to look for hints of that serenity, questions about oysters pressed less and fell away.

MASS STIMULATION

Brood-stock oysters, from which Justin would source sperm and eggs, waiting to spawn in tanks of water.

Justin spent the summer killing oysters. He spawned them from his precious brood stock, then they died. He spawned again and again, through May and July, and the oysters kept dying. Perhaps the water in the upwellers was too acidic; maybe they overfed on algae. He wasn't sure. Three months of death and questions left him undeterred. I returned to the shellfish lab in early August to watch yet another round, and I hoped I would bring Justin some luck. He didn't think he needed it. It was a process, something to which he was acclimating himself. He had been confident earlier in the year; following a string of failures, he had resigned himself to calling this a learning experience. What else could he do? With me as a witness, he might kill some more.

Throughout the spring, Justin had metamorphosed the lab that once stunk of bleach and mildew into a room of upwellers and downwellers, overhead and underfoot pipes, rushing river water splashing into tubs and then draining out and returning to the Skidaway, and a hint of bleach. This was his lair, a hatchery of his own making and design. The minutiae of spawning eased his mind, kept it busy so that his muscles didn't rebel. To calm his body, he ran daily with Rob Hein, heading two miles inland toward the Landings, a sprawling gated golf community on Skidaway, before heading back. The spawning issues he faced weren't headaches, just hiccups. He emailed leaders in the hatchery field, combed through reams of scientific journals. He settled on pH as the culprit, and likely the feed-rate ratios were off.

Spawning tanks—shoebox-sized plastic bins arranged like vertebrae under a spine of water-delivering PVC—were the only addition to the hatchery that I didn't recognize. There were more than two dozen small tanks in the system, but Justin had placed brood-stock oysters in only nineteen of them as a trial. He turned the red nozzle aimed down at each bin to let out a rush of Skidaway water. The oyster lay on the bottom, engulfed in liquid.

We inspected the upweller. I submerged my arm in the pool and scooped up what looked like a handful of sand. Baby oysters glistened in my palm under a wash of fluorescent light. Their tiny bodies, newly viable, were attached to micro cultch, a mash of crushed oyster shell used in hatcheries as a medium for spawned oysters to set on. As they grew, their shells would subsume the cultch. Ready for the next spawn, the downwellers were drained empty. I used the wooden frame that the tub sat on to prop myself up and peer inside. It smelled of the Joy liquid soap Justin used to scrub the vats between uses. Justin and Tom had leaned on an extension colleague from another university to help design the hatchery and order the right components. Bill Walton, the oyster guru at Auburn University, stationed at the school's marine science campus on Dauphin Island, Alabama, had visited Skidaway to offer his advice. When I told Justin that I had met Bill and toured the Dauphin Island lab, he was envious. "Man, I hope you took good notes," he said.

Unlike Justin and Tom, Bill didn't expend too much effort in enticing old-fashioned oyster tongers to become oyster farmers. He told them what was what. They balked, and that was that. Instead, he sought out shrimpers. Already dominant players in the Gulf, shrimpers had capital and heads for business, Bill told me. But someone didn't need marine experience to become an oyster farmer. "Some of the best shellfish farmers are plumbers," he said.

Although he outpaced Justin, career-wise, by a good twelve years, I saw many similarities between the two men. Bill had no interest in stopping traditional tonging in the Mobile Bay area, where his lab was located and most of the new oyster farmers lived. He hoped both cohorts could pursue their businesses in harmony. He was a scientist, of course, but his greatest thrills weren't measured in the laboratory. He wanted to see more people making a living off the water and doing so sustainably. His reverence for working watermen matched Justin's. I found the quality admirable in both men. Perhaps the biggest difference between the two was that Bill and the Auburn lab operated at a level Justin and Tom's lab had yet to reach. On Skidaway Island, they were still drawing watermen into the fold. On Dauphin Island, Bill was affecting lives. "Oyster farming is not done by wealthy people, so anything I can do that can reduce their expenses is tempting," he said.

I told Justin about the economic opportunity zone that Bill had created—a sixty-acre area in the middle of an Alabama bay where dozens of oyster farmers plied their trade under the guiding watch of Bill and his employees. They didn't have traditional leases, I told him. Anybody who wanted to farm just took a course with Bill, bought their gear, then rented out space in the oyster park at $250 an acre. I witnessed about a dozen farms growing oysters on hanging racks. Some employed pontoon boats with oyster tumblers stationed on deck. Without heading to land, the farmers could sort and clean their crop while on the water.

The only competitive advantage that farmers in the park possessed came from how many oysters they wanted to grow and how well the oysters were marketed. But cool as I found the Auburn oyster grounds to be, something similar didn't seem possible in Georgia. The Gulf didn't share Georgia's seven-foot tides, and most of the sportfishermen hunted out in deep water. Can you think of a low-traffic deepwater spot like that? I asked. He couldn't. The estuaries in McIntosh and Camden were certainly fertile, but the intricate systems of creeks and marshes, along with the large tides, that made it such an inviting habitat for creatures of all kinds made standard aquaculture practices difficult to apply out of the box. In a discussion with Dom Guadagnoli about Alabama's special use oyster areas, Dom told me that such zones were possible to implement in Georgia. To create one for commercial purposes would require extensive permitting and the rewriting of policy, something that would require a time investment for bureaucrats and managers like him, and patience on the part of watermen. An oyster park was a possibility, Dom guessed, but he didn't see a consensus building in the GSGA for one.

Justin and I leaned on the cinder blocks and pressure-treated wood used to buttress hatchery equipment and discussed how long the spawn might take. It was eight thirty in the morning. Some of the spawns that had resulted in no living oysters had taken as long as twelve hours to draw enough eggs and sperm to produce a decent bunch of spat. This could run all day.

Tom walked in. He and Justin discussed the type of food given to the young oysters. Try a half-and-half blend of the T-iso algae strain, Tom suggested. "Hit it hard for the first two days and then back it off," he said.

"I see what you're saying," Justin said. "Well, it won't hurt anything."

After Tom left, I asked Justin how things had been going with his boss. He stays focused on his work, Justin said. Tom planned to tackle a PhD, a degree that would help return graduate assistants to the lab. With only a master's degree, Tom couldn't oversee student research. Being able to do so would increase the capacity of the lab to help fishermen and conduct research as it had when Randy Walker ran the place. It was part of Tom's long-term goals to have the lab bustling with a dozen or more young marine scientists.

Tom spent most days solving problems of a different sort, leaving Justin to himself, mostly. What kinds of problems? I asked. Like finding permanent funding for the hatchery. The grant that gave Justin his job and the watermen their oysters had another year left. If this was to last, they needed money. Tom and his boss, Mark Risse, thought it best to find a home for the hatchery in the general University of Georgia budget, which was passed by the state legislature every year. In the spring, some politicians, members of the university's board of regents, toured the hatchery. They liked what they saw.

"Anytime there's someone from the statehouse here, it's probably a good thing," Justin said.

What did they talk about? They wanted to know how it all worked, he said. John Pelli, the oysterman who lived down the street from Justin, had been there, and Joe Maley and Charlie, too, as well as some people from the Landings. Justin just answered the technical questions, he didn't want to talk too much or make any mistakes. Meetings like that weren't his thing. Brushing shoulders with powerbrokers, he left that to Tom. The visit made enough of an impression that the hatchery found room as a line item on the UGA budget, a request for $150,000 to keep the project alive another year. Whether that money would materialize was out of Tom and the university's control. The state House of Representatives and Senate would hash out the budget particulars the following spring.

Justin couldn't worry about where the money would come from. He had an important chicken-or-egg issue to parse: if not one single oyster he spawned in the hatchery survived, what good would another year of funding do?

Unnerved by the pressure, Justin was ready to begin another spawn. He selected a few oyster from the brood stock and took them into a side room off the hatchery. Screens of varying sieves hung on the walls, their densities marked on the side in black: seventy-five millimeters, fifty millimeters. Boxes of Gold Seal Micro Slides sat between two microscopes on a steel table. He dropped the oysters next to a sink. He shucked one with an oyster knife, then he picked up a scalpel. To make a lot of oysters, he had to kill a few oysters, Justin joked as he cut into the oyster's gonad, using the blade to scrape out a milky substance into a beaker. (I reserved a retort that pointed out how many he had already killed.)

He sucked part of the extracted gonad into a syringe, squirted a sample onto a slide, and slid it onto the microscope stage. It was impossible to positively assign an oyster's sex with the naked eye. Under magnification, Justin could differentiate egg and sperm quite easily. Eggs were clearly visible, shaped just like an egg. Sperm evaded sight, appearing as light shimmers through the eyepiece. Justin wanted eggs. He would mix them with warm water and spurt a dose of the slurry into the spawning tanks. It worked like pheromones, he said, a natural kick start for procreation.

"Blast it, you men!" Justin howled. "Another dude!" All the funny things Justin thought in silence while spawning came out with me around. He had cut into three oysters already, all males. He shucked another one. "This one might be a lady. She looks gummy."

He was right. Armed with a beaker of egg fluid, Justin began the spawn. He unleashed a flow of warm water, about eighty-five degrees, into the oyster tanks, the excess splashing over the bin tops and down into a drain. Mimicking the warmth of spring tides, the stream should have sparked the oysters to erupt with milt and ovum. But they did nothing. He cooled the water down closer to seventy-five degrees, then warmed it up again, playing moon and sun to fake spring temperature fluctuations.

One male spawned, but it was nothing to celebrate. There would be plenty of males, always, Justin said. He sought a 60 percent female spawn. When the male clouded its tank with a pearly film, Justin stretched a piece of tape on the bin's front, just below its assigned number, and wrote down the time: 9:53 a.m. Then, more waiting.

"You guys aren't giving me anything today, nothing whatsoever," he said. "The wait can be nine hours, just sitting here watching. But once a female goes, they'll all go. I don't know if they pick up on vibrations or what."

He shook his head but remained chipper. He wasn't worried. Some of the whine might have been for show, since he had a captive audience. If so, I appreciated the performance. We had become friends, and he lowered his guard around me; he felt comfortable sharing the idiosyncrasies of his hatchery life.

Gone was the swagger of the Spatking; the grad school oyster nerd that Amelia fell in love with had returned. While I missed the masculine bluster, I enjoyed how relaxed he now acted in my presence, bad jokes and all. He called off the brood stock by nicknames: Captain Milky for the first male, Black and Tan for a dark-colored oyster.

Mass-stimulation time had arrived. Justin dipped a syringe into the egg beaker. It filled with white fluid. He dunked the syringe into the spawn tank and squirted in one or two puffs of egg. It's like a spritz of perfume on date night, Justin said. It should get them excited.

At 11:55 a.m., number 6 began to shake. Another male: a stream of sperm flooded the container. In a few minutes, it became hard to see the oyster through the seminal fog. He removed the spent males, then strained the sperm water through a sieve and into a bucket to remove waste.

At 2:30 p.m., Justin ran another round of cold water, followed by a warm one. It was the third cycle of the day, and no females had appeared. We reminisced about his time as a commercial oysterman. I asked him to tell again about his brushes with death. He didn't mind, since they were his favorite memories from his waterman tenure. The near drowning in Jones Hammock Creek in a thunderstorm. The gashes on his legs from abrading oysters. He recalled it all fondly.

"Do you think you'll ever go back?" I asked him, thinking of our conversation with Amelia over dinner. Justin applied more egg wash to the spawn tanks, took a breath, and responded.

"I'd like to, but I don't think I'll ever have a lease again in Georgia."

Why not?

"I gave mine up," he said. "Why would they give it back? I have what they want, the knowledge, and I still have commercial property," he said. He still owned his lot in Thunderbolt. "But I don't have the capital, and that's huge. And besides, they need people who live near the leases. I couldn't keep a good eye on my oysters living all the way up here."

Having been shoved down memory lane, Justin quieted. He patrolled the spawn tanks, taking a break to lean on the upweller and sip from an energy drink. He preferred to drink his calories on workdays. I bought us some of his favorite snacks from the closest grocery store—canned octopus, sushi, crackers—but he hardly touched the spread. Unlike the sleep-deprived, overcaffeinated Justin who had spent the afternoon with Rafe nearly a year ago, he fought off jitters. I sat down on the concrete floor, misted with hatchery overspray. The condensation began marinating the seat of my shorts. I compiled notes as Justin darted his attention between tanks, looking for the next twitch.

When the first female popped, at 2:47 p.m., and then the second flared out four minutes later, Justin, finally, was in business. Thirty minutes later, the third shot a thin stream of eggs out from its shell. Unlike the scattershot blast of the males, the eggs hung in the water like a string of crystals. He screened the females' water as he had done with the males. He took a sample and examined it under the microscope. He ran the numbers, estimated that he collected six million eggs from the spawn. He counted to assess how much sperm he needed to fertilize the batch. Not every egg would produce an oyster, he said, but many would.

He stirred up the bucket of sperm and dumped in the eggs. With a device like a plunger, he disturbed the liquid, enticing the two to meet. On another sample under the microscope, he looked for cell division. Convinced the coupling had occurred, that larvae had formed, he dumped the mixture into one of the giant green tubs where filtered and conditioned water circulated. Later, at two weeks of age, the oyster larvae would develop an eye and a foot. At that point, Justin would drain the tank, move the larvae into a set tank, and introduce microcultch. The larvae would set on the microcultch—ground-up oyster shell—and soon begin to resemble oysters as we know them. If they reached one millimeter in size, Justin considered them viable and the spawn successful.

We closed down the hatchery and left, going our separate ways. Justin appeared happy about the day's progress. It had taken most of the day, but his babies had lived up to his expectations.

STUMBLES

The wall of Justin Manley's office in the Shellfish Research Laboratory on Skidaway Island.

In his upstairs office in the shellfish lab, Justin stared at a computer screen. A black bookshelf stocked with clothbound academic journals stood against one wall to the left. Sunrays blared through a lone windowpane and softened the harsh ricochet of institutional light against white cinder block. On the wall, a map of St. Catherine's Sound was pinned to a corkboard, along with a pen drawing of a possible floating oyster cage setup; a marker drawing by his son, Mercer, of letters, eyeballs, and targets; the number for a Whole Foods seafood buyer; and a construction-paper heart with crayon writing stating, "Best Dad Ever."

On a November morning in 2015, he had been leisurely reading up on mangrove oysters, a new fascination. While on a recent Caribbean vacation, he had encountered bivalve mollusks growing on mangrove roots, attaching themselves to bark as they did on Justin's spat sticks. If climate change increased the already balmy subtropical Georgia temperatures, he joked, we might start to see mangrove trees migrate here. Kidding aside, mangrove oysters offered a new, creative area of exploration.

Talk of mangrove forest incursions marked the first time Justin had mentioned climate change in my presence as something he was seriously considering. The topic hadn't come up in the field, either. Environmental protections, of course, were ongoing pursuits of the likes of Charlie Phillips and Dan DeGuire. But climate change came up only in discussions of drought—specifically, when large rains following extended rainless periods flushed too much freshwater into the estuary, killing oysters. Even then, the politicized term *climate change* was avoided.

Despite the reluctance to discuss climate change, its effects were becoming more familiar to coastal residents. Perhaps the most extreme encounter: the increased frequency of road-closing and home-ruining floods in Savannah and on Tybee Island. In other departments of the Marine Extension Service, Justin's colleagues worked with residents to prepare for the inevitability of more floods, bigger storm surges, and the reality of living with higher sea levels.

Another climate change effect: marsh disappeared at an increasing clip, best noticed from outer space. In the summer of 2016, Creighton University researchers published an examination of twenty-eight years of satellite images taken of the marshes fanning out from the mouth of the Altamaha River in Glynn and McIntosh Counties. They found that between 1984 and 2011, spartina had died off by 30 percent. The culprit? Drought, mostly.

Scientists had begun to link the warming climate, and accompanying warming oceans, to a rise in food-poisoning cases caused by *Vibrio*, the sickening bacteria easily transmitted by raw shellfish plucked from too-warm waters. So far, the increased risk had shown up only in North Atlantic and North Sea samples, but if the data held up, and were scaled to other bodies of water, the effects on the industry as a whole would certainly be negative.

Other scientists warned that ocean acidification threatened to ruin North American oyster populations, and had already wreaked havoc among West Coast farms. Ocean water is basic. As seawater absorbs increasing amounts of carbon dioxide, it pushes pH levels toward neutral, which halts an oyster's ability to produce a shell. I discussed the issue with Brian Hopkinson, a professor in UGA's Marine Science Department whose research focuses on the topic. In an email, he wrote that he didn't expect acidification to become a big problem along the Georgia coast. The wetlands here already produced high levels of CO_2 from natural decomposition processes occurring when water was flushed from rivers into the estuary and out into the ocean, as well as from the rot of spartina into the mud. Georgia oysters, therefore, were especially prepared to withstand acidification.

Later, when Justin and I discussed the effects of rising seas on oysters, farmed or wild, he reserved judgment. Lots of variables, so much we don't know, he said. Increased salinity as seawater overrode the ratio of freshwater to salt water in the estuary and on into the marsh would surely affect oyster survival.

The loss of oyster beds, and their millions of filter feeders, would mean gunky wetlands. Without oysters to process algal blooms from the water, marsh waters would drastically change, the water that fish and crustaceans swam in would drastically change. Perhaps more than any other wetland denizen, the oyster is the ecosystem's linchpin, which is why its survival mattered. Oyster death en masse wouldn't be a canary warning us to act fast before disaster struck. If they died, it would already be too late.

With hopeful eyes, it was easy to gaze out over Georgia's robust wild oyster beds and trust in their resiliency. Oysters are adaptive creatures: there was no reason to believe the animals couldn't build taller beds or move inland as the Atlantic Ocean invaded St. Catherine's Sound. Still, such thoughts were prospects, not guarantees. Clear answers were fleeting, but undeniably, something was happening. Given the uncertainty of aquaculture in the face of climate change, Justin concentrated more on tasks at hand, keeping climate change "always in the back of my mind."

Fall and winter were the hatchery's slow seasons. Rob kept busy out in the field, delivering seed and gear, running small grant-funded experiments on fish populations or water quality, just as Justin used to do. Stuck in the office, preparing for the spring spawn, Justin contended with downtime. It made him "cerebral," he said, which I construed as philosophical, reflective.

Did he feel like he was adapting to life in the office?

"I'm progressing," he said. "Before, I was just helping people solve problems. Now I'm managing and planning. I'm creating something. It feels awkward to be off the water. I've got this cabin fever, like spring fever, but for the oyster season in winter. I want to be out there, but I can't. I have this thing to do here. Anyway, I've been hard on my body. I'm getting old and it's catching up."

He spoke of arthritis, of damage that began years ago back in Detroit, back in the auto factories of his father and grandfathers. The physical weathering of his oystering life wasn't relegated to just his body. The *Green Hornet*'s motor had died. No use pouring money into fixing it. To him, it signified a symbolic untethering to the thought of ever having another oyster business again. "That boat almost got me killed a bunch of times," he reflected. To be fair, it saved his life plenty of times, too.

Was he finding any moments to interact with his old colleagues?

Just through a little "verbal information exchange." But even that was best handled via Rob, who, he said, was settling well into the job—marine extension agent—that Justin often missed performing. Rob is smart, he said, a fast learner, and has an ability to see the bigger picture. He lived in a dorm room on the Skidaway campus. With only five TV stations to distract him, he read furiously. He had brought himself up to speed on aquaculture and was readily applying what he had learned. Within institutions such as the Marine Extension Service, and science in general, it is always helpful to inject vitality in the form of youth. Rob's fresh ideas were welcome, Justin said.

How about the oysters from the summer spawn? How had they fared?

Justin laughed: every egg he captured that August day had died.

"We successfully spawned five million eggs from one female that I worked with for seven hours. We murdered every larva," he said. Luckily, he tried again a week later. The larvae survived, and he was able to grow enough viable oysters to deliver another round of seed to the growers. It was an eleventh-hour success.

Rob poked his head into Justin's office. He secured his hands onto the doorframe; alight with energy, his muscular physique bobbed, uncontrollably eager. With curls bleached by the sun, wearing a tank top and board shorts, Rob exuded the qualities of two compelling male archetypes: the southern country boy and the surfer bum. He was whip smart, buoyant, and ruggedly handsome.

"You ready?" he asked Justin.

"Let me get my shoes," he responded.

It was time for their daily miles—a half-hour run up McWhorter Drive, then back to the lab. Got to stay in shape, Justin laughed. Got to hold this old body together. Rob inspired him to keep it up. As they eased into a pace that would last them four to five miles, I said I would catch up with them tomorrow.

Rob had helped me organize a boat trip for a photographer friend, a woman named Rinne Allen, who wanted to document a wild oyster harvest. I asked John Pelli, the oysterman and clammer who lived just up the street from Justin, whether we could visit him during a harvest. He said sure, but we would have to get ourselves to his site on the Half Moon River, about a forty-minute boat ride from Skidaway. Rob offered to give Rinne and I a ride.

The three of us sat in a university boat, waiting for John's skiff to rush by toward Wilmington Island. Rob pointed toward a spot near a bend in the river. He had been growing experimental oysters there, trying new gear that he had read about, and he was impatient to get them ready

John Pelli.

for Georgia's nascent growers. Echoing the belt-tightening goals of Bill Walton, Rob boasted about the mariculture tricks he had learned.

"I've learned to farm a pretty good oyster," he said. "I can baby them, pamper them, clean them all the time, and I've got really expensive gear. What I have to do is make what I've learned work economically for them."

John's wake interrupted our chat, and we rushed after him in chase. Soon, we joined our boats up a tidal river off the Wassaw Sound. On a beautiful November evening, nearing dusk, the fading sun clawed at the edges of an ashen sky. Its setting tore tangerine-colored streaks on the horizon line. The brooding eventide gave the spartina's green hue the luminance of neon. But the river, stirred by wave action and an oncoming storm, stewed in darkness. Against this contrasting landscape, John's team plucked oysters from the Half Moon's bank with deliberate speed. Five bushels before sundown, get on it. As Rinne snapped her pictures and I chewed John's ear off, Rob donned gloves, grabbed a piece of steel, and joined the harvesting fray. As they worked, Rob and John talked about the possible aquaculture gear that Marine Extension might be able to offer.

"Man, I really want floating cages, that would be great," John Pelli said. "That's my main concern, getting wet. I want to be able to work at high tide. Most days I love this, it's the best job in the world, but when it's raining and I'm soaked, I can doubt why I ever wanted to do this."

"It'll all depend on the site. I'm happy to come back, and we can look at sites together, John. That's a big part of my job," Rob said. I could tell Rob had taken extension's mission of service to heart. Or perhaps he hoped I would report back positively to Tom, but I doubted that. He was honest, hardworking, energetic, perfect for this kind of work. But I knew he wasn't long for this post. He told me he had been considering a move out west to learn aquaculture on a massive scale in Washington State.

When we returned to the Skidaway dock under the cloak of night, I asked Rob what he thought of Justin, what the differences between the two of them were.

"Justin had to go out and bust his ass all those years," Rob said. "He had to prove himself. I think I've had an easier time getting acquainted with the guys than Justin did. I grew up here. I can put a dip in my mouth and talk with a really southern accent if I need to."

I joined Justin and Amelia for dinner that night. Just off the water with Rob and Rinne, I arrived late. The house lights were dim, the kids were already shuffling toward bed. We ate with tired appetites, happy to be in one another's company.

Lying on a side table, I found a picture of a younger Justin. Late high school, maybe early college. He wore a tuxedo and appeared to be exiting a limousine, flanked by two young men also formally dressed. He said senior prom, which made sense. What stood out, or up, was his hair. Gelled and groomed into a high-top fade, the coif resembled the one on the early 1990s rapper Vanilla Ice. Like a block of polished beech, the hair appeared immovable. I wondered whether all the product he used in the 1990s had caused all that hair to fall out. He laughed. "That was a long time ago."

I questioned Justin further about these mangrove oysters. Was his interest in them real or a fleeting curiosity? The latter, it seemed, since he admitted he and Amelia had been daydreaming about a move to the Florida Keys, where he first saw a mangrove up close. Amelia's skills would allow her to find work almost anywhere—people of her medical caliber were rarely unemployed. They were entertaining the idea of a tropical life. I told him what Rob said about moving to Washington to work on a major oyster farm.

"That would be an awesome experience. It would be hard for me to leave the Southeast after all this time," he said.

Amelia interrupted him: "We should move, because what has the Southeast done for you?"

I didn't know what to make of her frustration. Was this about the failure of the Spatking? Couldn't be. What would happen when the grant funding dried up? Amelia worried that all of Justin's effort wouldn't pay off, but not for any lack of effort on his part. Would the state permanently prop up the hatchery and her husband's position? A hunch told her the answer might be no. If the powers that be couldn't support the hatchery work, if they couldn't promise them a future, maybe they should begin to make other plans.

Justin felt differently. As he had said in an earlier conversation that week, he was building something. He wanted to see it through, even though the payoff might not be what he was promised. He might cast a fantasy toward warm Caribbean waters and wish himself sunburned and liver-full of rum drinks, but he had committed himself to the Georgia oyster industry. His fight still had focus.

If he was pensive these days—"cerebral"—it was because the failures of the summer spawn had quieted him. There were flaws, and they were visible. The weakness the plan showed in its early stages had drawn the ire of an important waterman, and Justin took it personally. In the morning, I planned to drive to hear the complaint in person, at the desk of Charlie Phillips.

At the stop sign at Greystone Road, a half mile from Sapelo Sea Farms, a truck trailing a boat left the public dock accessing the Sapelo River. I was headed to the same place. The truck turned, and I caught sight of the driver—Mike Townsend—and we locked eyes. Months had passed since we last saw each other, but he recognized me in an instant.

"Hey, man," I heard him shout through the wind. He pulled off onto a grassy shoulder. I turned my car around and parked in front of him.

Rain misted us, part of a front that didn't seem too threatening, according to weather reports. Clouds looked like an airship fleet in the sky: heavy as steel and ominously low—an odd partner for light precipitation.

"What are you doing?" Mike asked, leaning on his truck.

"I'm supposed to meet Charlie after lunch. Just killing some time," I said. "How did your oysters do?"

"Oh, I think about half of them died, but I didn't tend to them all that much. I heard that was about the average, though. Is that right?" he asked.

Justin and I had discussed this very topic the day before. Most growers in the aquaculture project had lost about 50 percent of the oysters given to them by Marine Extension. The losses did not cause Justin any dismay. "That's pretty good," he had said. "And it's still a good return for the guys."

Mike sounded unperturbed. His personal investment was low. When I asked him how things were going in general, oysters weren't at the top of his mind: "It hasn't been that good a year for us cast-net shrimpers," he said. "Haven't been too many of them out there for us to catch."

Charlie Phillips, a waterman with a great deal invested in shellfish, noted that high mortality. And I was about to get an earful about it.

Mike showed me his cheerful side. He was happy, daydreaming of a sailboat trip to Bermuda if he ever found that sugar momma. He saved his cynical, anti-Charlie snipes until we parted ways.

"You watch out for Charlie, now," Mike said, laughing.

I continued on to the dock and walked out to the edge of the gangway. The Sapelo River rumbled in silence, low-breaking wave caps looked as if they had been dipped in mercury. A silver haze covered the marsh and distant barrier islands, subdued the wetlands' repetitive green hue. I had months still left to travel along the coast, but I had already begun to miss this place. People who grew up next to the ocean, as I had, find true calm in salty air. In this gray winter brume, a mask of salt spray exfoliated my skin. Waves toppled and flattened in a hypnotic pattern. I inhaled the biggest breath I could muster, held it for a moment before release, then headed to Charlie's.

He had been busy, of course. That was his nature—what else could he do? He had taken over the Pelican Point restaurant from his father. He envisioned its renovation as a dock-to-table restaurant, with the dock within sight of the dining room. Perhaps he would install teaching tanks in the waiting area so that people could see how oysters filtered water. He wanted to make it clear that the fresh seafood people ate in this restaurant—which he planned to call Fish Dock—was directly affected by what happened upstream. If you liked that food, Charlie planned to tell future diners, you better watch what you put on that lawn.

Sounded like a ton of work for someone so busy, I said.

"Like I have a bunch of spare time anyway," Charlie said. "At my age I'm supposed to be downsizing, but I'm putting a damn coal on the fire."

I told him I had seen Sapelo oysters offered at a restaurant where I lived. It was a rarity, and a cool one. Charlie had picked up another oyster lease in the past year. He was buying plenty of oysters from the McIntoshes and from Jeff Erickson, selling them up the coast and inland as Sapelo oysters. Overall, wild oyster sales were up, he said.

"A lot of people are finally figuring out how good these local oysters are, and I'm starting to put them at restaurants and events. People are getting used to the—you think they would know it by now—but maybe it was just a matter of getting it in some places. Historically, it's just been oyster roast stuff. But I've even been sending my oysters up to the Landings, they're enjoying the salinity of them. And I think if the Gulf keeps having these problems, and we keep having these water wars, we're going to see this continue."

One of Charlie's young workers walked in, asking for a clarification of orders from earlier in the day. He was a thin, fair-skinned kid, his skin burnt all to hell. Reminded of a story that Charlie once told about me about a card-playing employee who took a day off work after being shot, I examined his body for a gunshot scar. I couldn't find one.

There had been losses with the farmed oysters, Charlie continued, more than he was comfortable with. A reminder of it spun him into a tizzy.

"They had a heck of a time up there. I don't know if they've got it figured out yet or not," he said. "The way they're running the rabbit—between UGA and the Coastal Resources Division—they don't want you to order seed from anywhere else, they don't want to order triploids, they limit the way you can grow them—there's too many choke points for me to think it's viable. And I can't keep up with clams. It's hard for me to consider doing it with so many choke points. If I could play around with triploids, I'd be all over it. I don't see where it's going to be economically viable. The industry needs some latitude where you can try stuff, just like when we built the clam industry."

Charlie was talking about triploid oysters, a lab-born bivalve that didn't spawn—and therefore could be eaten year-round—and plumped to market weight in half the time as a regular oyster. Hatcheries in Virginia sold them, as did ones in the Northeast and Washington. Triploids were common in most oyster aquaculture in the United States, but Dom, the GSGA, Tom, and Justin had urged that they not be allowed in Georgia, at least until a triploid could be produced in the state.

Oysters have two sets of chromosomes, making them diploids, just like humans, with one set of chromosomes from each parent. Triploids have three sets. The rare triploid—maybe one in a million—does spawn, but the majority don't. Because the triploid isn't expending energy on filling its gonad with eggs or sperm, it grows fatter faster. A triploid gets meaty enough to slurp in nine months; a wild Georgia oyster takes three years, on average, to become three inches long, and even then its muscle, unless the oyster is harvested and eaten in the spring, when its about to spawn, looks rather limp compared to that of its triploid colleague. Triploids, or "trips," do occur naturally—many widely consumed crops, like potatoes, peanuts, and wheat, have more than two sets of chromosomes—but in oysters, they are a rarity. The mass-produced commercial trips that account for most raw bar menu options come from a hatchery, never nature. Often, the trip is the child of a tetraploid-diploid spawn. A tetraploid—an oyster with four sets of chromosomes—is the result of decades of painstaking lab research that drew influence from salmon husbandry. Without the invention of tetraploids, triploid breeding would be impossible. Therefore, patents for tetraploid production are closely held, and these trip parents, usually males, are kept under tight hatchery security. Without a tetraploid in the breeding process, a hatchery can't effectively spawn triploids. Some reactionaries have referred to trips and tetraploids as Frankenstein oysters—one article called them "tasty mutants"—but there was nothing to fear: this wasn't genetic manipulation, and there was no risk to the environment or our health.

Justin and Dom knew triploids poised no public threat. But as they had discussed at that GSGA meeting in June 2014, outside seed wouldn't make its way into Georgia, at least not as immediately as Charlie would like. Charlie had gone along with it at first; maybe he hadn't predicted how popular local oysters would become. With a little research and a phone call or two to commercial shellfishing buddies in other states, Charlie could surmise that imported triploids would put Georgia oysters on the market quickly. He wasn't wrong. What Charlie actually bristled against was Dom and Tom's take-it-slow approach. He didn't think this hatchery project benefited the industry, at least not in the way they were running it now. "The numbers make them look good," he said. "They don't account for labor and cost of labor."

He repeated the term *choke point* again and again. There were too many choke points in farming oysters by following the rules laid out by the DNR and Marine Extension. If they removed some choke points, maybe he would feel better about it all. I reminded him that Justin and Tom were hoping to put themselves out of business; they thought a commercial hatchery would open, especially if enough growers succeeded in this pilot program.

"They say they want a commercial hatchery to come in? Great, there's another choke point," Charlie said. "What if they jack the price up? At my age, if I don't have something clicking in say, five years, I don't know if I'm going to be around that long."

I could see now he was determined to be a curmudgeon, no budging. It wasn't my job to change his mind, either, so I let the man ramble. He had no interest in playing a long game on oysters; that was fine. If so, what did the future hold for Charlie? More fishing boats, of course. He had to feed that addiction. New catches, too. He hadn't explored octopus at all. He had built something, but had no one to pass it off to. How did he see this all playing out? I asked. He would die at work, he said. He hoped to be pulled from behind his desk. Better yet, dragged off his beloved airboat. "I hope I just fell over," he said.

"How mad was Charlie at us last time you talked to him?" Justin asked when I saw him. Justin, Tom Bliss, and I were standing around a tank of algae—feed for larval oysters—outside the Marine Extension lab and talking shellfish politics.

"He's pretty angry about the triploids, and he doesn't think you know what you're doing," I said. The GSGA had voted a few years earlier to not allow imports into the state, which also meant triploids. Technically, it had been a unanimous vote, but Charlie had never been for it, Justin told me. He had to be prompted to go along with the group.

"This was when you were still with the group, right? How much of this was your doing?" I asked Justin.

"I always tried to be objective as possible, but I have an affinity for these oysters," he said. Importing oysters from the Gulf or the Northeast would be like playing Russian roulette, he said: eventually something harmful would sneak in with a batch, as history has proved. We have the most to lose by bringing in out-of-state seed, not the hatcheries up north, Justin said.

It seemed like Charlie was throwing in the towel on the oyster experiment. Neither Justin nor Tom believed that, and I wagered then that his fussing with me could have been smoke and mirrors. Charlie had no interest in putting sweat equity into oyster farming, they said. He wanted to be able to apply the absolute least amount of labor to meet his margins. What he wanted, Justin and Tom thought, was to buy up every seed the Skidaway hatchery could produce and sell whatever he could with the smallest input of effort. He wanted seed to be plentiful and cheap. It was how he had made a killing in clams, they said. Problem was, oysters are fickle animals compared to clams. To survive in aquaculture, they required attention, while clams liked sitting in the mud.

If Charlie could buy seed from an out-of-state hatchery, he could outpurchase all the watermen, undersell them, and try to control the market. It was just how his brain worked, they said. Charlie was a necessary tempest. The industry needed an alpha, and Charlie ruled the pound. He enjoyed bending the ears of bureaucrats, politicians, and community leaders. That power could be harnessed for good, but the fear that he might bite his pack mates was ever present.

It all sounded crazy to me, putting up with an angry Charlie as they did, but everyone, including Justin and Tom, but not Mike Townsend, were used to it. They didn't mind the drama.

"I appreciate his frankness," Tom said. "Some of the guys aren't forthcoming, but with Charlie, he tells you what he's thinking. Sometimes it makes me furious, sometimes it makes me really happy."

COASTAL DAY AT THE CAPITOL

A table at Coastal Day at the Capitol, Atlanta.

The way Joe Maley described the Department of Natural Resources' Coastal Day at the Capitol in Atlanta made it sound like a can't-miss event. It wasn't likely that I would have as quotable an experience as Joe had had in previous years, but the thought of watching salty old oystermen network with pleated and cufflinked politicians seemed so perfectly diametric that I had to be there. I called Dom, who organized the event, to ask permission to show up. Of course, happy to have me, he wrote via email. He looked forward to finding an opportunity to talk face-to-face. I called Justin to ask whether there was any way he might be able to make it. He would have to see, he said. He would have to ask Tom.

Some protocols around Marine Extension had changed, Justin told me. It wasn't anyone in the shellfish lab's fault, but the leash on Skidaway employees had been drawn tight. The basic rule was straightforward: nobody was to talk to the press before consulting higher-ups, Justin had been informed. It wasn't Tom's decree; the concern over employee blabbermouthing came from many links up the chain of command, nowhere near Skidaway. But if a reporter—or anyone, really—called Justin for a comment or any kind of information, Justin was supposed to forward the request along to Tom. Even if Tom eventually sourced the answer or comment from Justin, the process had to be followed.

On my last trip to the lab, Justin told me to announce my presence in advance via email to Tom, who would then copy the requisite managers on a forwarded email, making them aware of the nature of my visit. I felt that neither Tom nor his bosses wanted to keep me from my research, but I realized how lucky I had been to have well over a year of unfettered access to Justin and the inner workings of Marine Extension.

A week later, a few days before Coastal Day, I called Justin back. The voice on the other end of the line quaked. He was frustrated. Tom had told him that if it was necessary for someone from UGA to be present at Coastal Day, Tom would have do it himself. The instructions hewed to the strict code of conduct that had been established—if a public face of the shellfish lab was required, Tom would fill that role. It was the response Justin expected. Still, the decision surprised him.

Administrators had muzzled him, he growled. Why quiet the guy who knew the science? Why did he have to email three people in order to talk to a reporter? They were strapping a shock collar around a wolf's neck then wondering why he was pissed, Justin said. He was trying to fit in: they wanted a domesticated animal, but a wild streak still ran through him. He clawed at the lab's boundaries. Justin's photograph had appeared in magazines. Journalists, including myself, stationed him as an authority on the rising prominence of southern oysters. Now his voice had been quieted.

The more we talked, the more Justin calmed. He desperately needed to vent to someone other than Amelia. The root of his frustration wasn't Tom, or that he couldn't speak his mind, although that was certainly annoying. He felt cut off. No more fieldwork, no site visits; he missed the chatter with watermen. "It's like I can't even see my old friends," he said. Watermen didn't visit him on Skidaway. To complete his mission—to launch a lab-hatched line of endemic oysters—he couldn't wander far. His duties necessitated his proximity. He had partly built this prison himself.

Long story short, he wouldn't be in Atlanta. But neither would Tom. In truth, Atlanta didn't matter that January. They had spring projects to prepare for.

Justin's outpouring supported a theory about him that I had been developing. He was passionate. His dreams were bigger than himself, and he exhibited the foresight to chart their realization. In the end, I wondered whether the freedom he gave up in accomplishing dreams was worth it.

Justin was a natural entrepreneur, but he was also gifted in the realms of science and research. To combine the two strengths didn't suggest disaster, but in Justin's case, each self competed against the other. To flourish, they would have to complement each other, or one would have to give up the fight. He could run the lab, but the lab was not where he needed to be, I wagered. He deserved a life on the water. It was what his nature demanded of him, and I had come to believe that the lab stifled his better self. His better self was the Spatking.

This intense moment on the phone, I thought, proved my inkling correct, though my confidence wouldn't last. In the next few months, Justin would reveal my theory about his true nature to be the selfish act of a storyteller. A wild Justin lashing at his cage served me better than a patient Justin acting in his long-term interests. I did no service to Justin the scientist, the husband, the father, or the autonomous spirit. Perhaps it was best if I stopped theorizing about the man altogether.

I had first sketched Justin as a man with rough edges—with a shade of unpredictability and highlights of masculine obstinacy. After all the talk we had shared about the nobility of watermen, here was Justin becoming their opposite. This wasn't a bad thing. It was part of a strategy.

This conversation highlighted that Justin's theatre of war had shifted. With one final harangue, the Spatking fully aimed his efforts elsewhere. He planned to be a good boy, mostly, from now on, to play for the team. By letting me witness a last howl, he showed me that there had been protest.

"I'll tell the guys hello for you," I said.

"Please do."

The Georgia State Capitol, capped in gold leaf, encompassed one grassy city block flanked on three sides by cement boxes. On an unseasonably sunny Thursday in January, the legislature was in session. Politicians, aides, and lobbyists scurried past cooks sautéing wild shrimp, onions, and peppers in a wide metal pan, then slipped inside the Capitol through security gates. Near the sizzle of aromatic crustaceans, a cauldron of grits bubbled. The cooks prepared the shrimp and grits as a Coastal Day treat for lawmakers, a culinary reminder to keep the coast in mind come budget time. As Joe Maley liked to say, politicians aren't the type to skip a free lunch. I asked the cooks where I might find some oystermen. They pointed to a window one floor up. Bottom floor of the rotunda, they said, a conference room to the left of the staircase. There's a sign.

It was 11:00 a.m. Upstairs, Joe, Dan DeGuire, and Danny Eller shucked furiously. They wore gloves on one hand, and wrapped a towel around the oyster as they plunged into the hinge with a knife grasped in the opposite fist. The gloves were made of cloth, not the fancy steel protectors that professional restaurant shuckers employ. A plastic layer on the palm helped keep the oyster from slipping from their grip, but wouldn't protect the skin underneath from a penetrating blade.

Dom had joined in, too. They placed the opened oysters on four metal serving trays laid out on a folding table. Underneath was a plastic sheet. They shucked on top of it, discarding shell tops into an empty cooler. Droplets of oyster liquor pooled on the table and spilled over onto the floor, joining shell shards on a black rubber mat. "Everybody still got ten fingers?" asked a gray-haired DNR administrator. The guys didn't find that very funny. They were very tired.

Dan had left Camden County at one in the morning to drive to Sunbury. Joe, Danny, and Dan drove to Atlanta in Joe's truck, leaving at four. They had brought with them five hundred oysters, harvested the afternoon before, in insulated coolers.

Outside the conference room, people perused informational displays about gun advocacy and land conservation. Soon, throngs of lawmakers and staff would be shuffling through the oystermen's room, filling plates with shrimp, thanks to the Wild Georgia Shrimp industry group, and oysters from the Georgia Shellfish Growers Association.

The oystermen were quiet, awaiting a storm of eaters that wouldn't let up for an hour. Dom set up sleeves of saltines, hot sauce bottles, and lemon wedges. The guys kept shucking. I stood behind them, my back against a display about aquatic life: pictures of fish, wetland ecosystem strata, and a wooden block depicting the lifecycle of an oyster, from spat to full shell, in etchings and images. The room had two doors: the eaters would enter through one to my rear and leave through the other one just next to the oyster table. A line appeared at the former. I peeked through the space between the door and the jamb and saw a line that bent and stretched so far that I couldn't see its end.

A woman with a DNR lanyard corralled the line. Once the shrimpers and the oystermen gave her the go-ahead, she released them to fill their bellies. I distinguished the VIPs by their name tags. Legislators wore rectangular badges that noted their post in the Senate or House, and the location of their constituency.

B. J. Pak, District 108, Lilburn, came through early. "You guys have the hard jobs," Pak said, eyeing the shuckers as he piled oysters onto a plastic plate. Danny, a ball cap pulled low over his brow, responded without meeting Pak's eyes: "Yeah, man," he said, the words labored in a deep whistle through a gap in his top teeth. Danny, I had been told, had struggled recently with his health. Diabetes kept him off the water most days. His nephews filled in where they could. He looked exhausted, but perhaps I'd mistaken his mellow surliness for fatigue. He was a fish out of water, after all, shucking these oysters for men and women with far more power and money than he. I didn't need to judge his mood or appearance.

Behind me, a woman commented on the life cycle display: "I'm not sure I wanted to see that," said Michelle Henson, District 86, Stone Mountain.

"You really don't want to know the life of an oyster before you eat them," said John Yates, District 73, Griffin.

Four men shucking couldn't keep up with demand. They had been going nonstop for an hour, but had hardly made it through the first of four bushels. Dan plunged two hands into an open bushel and extracted a bunch. His oyster knife cracked at the shells, making chink-chink sounds that were lost beneath the din of voices.

David Knight, District 130, Griffin, chairman of the Game, Fish, and Parks Committee and member of the Higher Education committee, announced that he had toured the Skidaway hatchery. "Is anything coming from that?" he asked the oystermen.

"Oh, yeah," Danny responded.

"Okay, I just hadn't heard anything," Knight said.

Joe leaned toward me to grab my attention. He pointed to a bearded man in a gray suit nearing us in the queue. He whispered that he was the guy who had asked last year about how the water wars were affecting Georgia oysters. Emory Dunahoo, District 30, Gillsville, smiled at Joe, clearly remembering him from last year. Joe certainly hadn't forgotten his face. They chatted as Dunahoo arranged oysters on his plate. He blotted the meat and liquor with dabs of Texas Pete hot sauce, a composition that looked like beads of blood dripping onto submerged rocks.

"When these guys call on you," Joe said—"these guys" meaning the DNR—"give them some money."

"I support y'all as much anyone, ask around," Dunahoo said, speaking over his right shoulder as he slipped out the door.

The DNR booked Coastal Day at the Capitol each year to remind lawmakers why the department deserved taxpayer funding. It was true that the DNR, in the past, had been hit by budget cuts. Fewer field agents prowled the state than in previous decades. As for Dom and the oystermen's concern, money to pay for federally mandated water testing dried up, forcing Dom to scurry to keep the program hobbling along. That money had also begun to trickle back.

What wasn't secure or funded beyond the near term was the hatchery. The original grant that gave Justin his job, employed Rob, and supplied watermen with gear and seed would reach its end in September. Beyond that, no money had appeared. Tom had conferred with his boss, Mark Risse, and they decided that the best approach was to find room in the state budget as part of the University of Georgia's yearly allocation. They hadn't asked for an outrageous amount. For a year, the hatchery wanted $150,000.

During the ensuing legislative session, the hatchery's fate would be decided, as would the question whether Justin and Rob would have jobs once the last month was ripped off the 2016 calendar. As a line item, the hatchery would have to survive House and Senate revisions before hitting Governor Nathan Deal's desk. With so much at stake, it seemed worth it for UGA to send a representative to Coastal Day. An extra handshake with a senator wouldn't have hurt. The men and women of the Gold Dome would have been able to taste and touch the difference for themselves. But I could see the argument for Tom and Justin staying home. A nuanced conversation about aquaculture would have been impossible at Coastal Day. Besides, it was DNR's day. No need to steal the spotlight.

More people came through the line. More lawmakers. More aides. More interns. More lobbyists. They filled their plates and disappeared through the door, some returning to an office to eat in peace, others mingling in the rotunda. It became clear that attendance for this coastal luncheon would be greater than expected. Only every third or fourth eater, it seemed, remembered to thank the guys.

Danny winced. His knuckles ached, he grew tired. He pulled at his fingers to alleviate the pain. He shook his hands.

"Want to hit the bench, Danny?" Dom said. "We can call in the second string."

"I'm fine," Danny said, stretching a glove back over his hand and picking up his knife.

The eyes of Ed Rynders, District 152, Albany, grew large at the plethora of bivalves before him. "Man, thanks, y'all," he said, slurping down a few before leaving the conference room with a plate full. "These are almost as good as..."

Joe's eyes flashed toward Rynders. He scowled and pointed the tip of the shucker toward the lawmaker: "I've got a knife!"

Rynders laughed: "Well, I do love Apalachicolas," he said.

"They're just rocks," Danny said.

Joe leaned in to me and bent my ear. "The hardest part of all this is just listening to all their witticisms," he said, shaking his head.

State senator Chuck Hufstetler asked for an oyster knife to shuck his own.

"We've got an EMT on standby," Joe joked.

"Yeah, but he's off eating oysters right now," Danny said.

Hufstetler's first attempt produced a dead oyster, a "muddy duddy" filled with gray muck. With another try, he found a beaut. He flashed a smile and exited.

Joe's Liberty County House rep, Al Williams, cut through the egress door, opting not to wait in line. He approached the table, and he and Joe lit up when they saw each other. The two men exchanged elbow daps. For Joe, this didn't constitute an endorsement of his elected official. They were just two old country boys trying to make good in the big city. Joe for a day, Al as a career. Al bellied up to the table and slurped down a few oysters. "Liberty County," he said. "Where the water tastes like cherry wine!"

"Oysters been dead for years, but now they're coming back, huh?" Representative Williams said to the group. Joe pointed his shucker at Dom: thank him, he said.

"Is that right?" Williams said.

"It's not just me," Dom said, blushing a bit. Al ate more oysters.

"Al, if you eat another one, I'll run against you," Joe said.

"Well, I'll quit," Al said, slapping his knee with a laugh.

The guys ran out of oysters. The shrimpers ran out of shrimp. The only grits left had hardened along the edge of hotel pans. The line thinned out, then emptied. Every other minute, someone would pop their head in and ask whether there were any left. Eventually, someone closed the door.

Dom stood off to the side of the room and fraternized with his DNR colleagues. Wondering what he hoped to talk to me about, as mentioned in our email exchange, I walked up to him and asked.

He wanted to make sure I was getting the correct impression of the changes the industry was going through. Industry rebuilding happens when all parties stick together, he said, and the watermen, UGA, and the DNR were all on the same page. They worked as a team.

"We have our disputes and vote different ways, but our goals and mission stay the same," he said. "That's how we succeed."

He said he had convinced his higher-ups to back him on splitting up leases. They were sold on the idea, but we have to make sure oyster farming is going to work. His waiting list for new leases grew longer by the day. As soon as it looked safe to open up the game, the boom would be on.

What disputes have you had? I asked.

Seed. Mostly seed, he said. This meant, without him saying the name, that Charlie Phillips's pro-import campaign was continuing. Thing was, Dom said, there had been a vote. It wasn't the DNR or UGA that decided to keep out imports. It was the fishermen themselves. He said it was important to note this because the industry had to advance with its blue-collar watermen at the forefront. He was determined to make sure the traditional methods of oystering remained an option for coastal residents.

"I get calls from out-of-state seed guys all the time, and they're like, 'We're safe, why won't you let us in?'"

All the stakeholders had discussed it, he told the prospective interlopers, and banning imports was how they had chosen to proceed. It was Dom's choice how to proceed, but he told me he was just following the watermen's lead. Dom didn't mind holding the door, but the pressure was mounting, and I suspected he hated having to say no all the time.

When we first met in his Brunswick office, he told me his approach was to wait and see. His tack hadn't changed. In the year since that meeting, he waited still. In the meantime, the tally of those kicking their heels had increased. In addition to the hopeful lessees and seed sellers, a couple of entrepreneurs in the Brunswick area had begun to bend Dom's ear. They were interested in opening a commercial hatchery, he said. Three or four businessmen, none of them with experience in aquaculture or shellfish propagation, mind you. Just sharks who could smell an opportunity. Before they bothered to write the business plan, though, they were waiting for the baseline data from the UGA experiment. As he said, the coast could burst open any moment with an aquaculture rush. Dom was determined to control the explosion. In growth of this kind, he said, it's important to not displace traditional fishermen, although there was no way he could know how likely that was to happen. He didn't want the fuse to light so fast that they were left covered in rubble.

That was it, he said. He only wanted to make sure he and I were on the same page. Everyone was working together for the best interests of the group, the coast, the future of oystering. Rising tides and whatnot.

Joe, Dan, and Danny packed up spent shells and trash into coolers, folded up the tables, rolled up their GSGA sign, an emblem reading "Semper Fi Ad Mare." Always faithful to the sea.

I followed them down the stairs and out through the security gate to a loading zone. It was lunchtime; the staffers who hadn't passed through the shrimp and oyster events hustled off in search of sub sandwiches or burritos. We piled all the gear near a curb, and Joe walked off to grab his truck. Dan and I talked about the fight to stop the Camden spaceport. He thought it was malarkey to place a jet-fuel-spewing launchpad belly up against some of the most pristine wetlands in North America. It reeked of corruption, Dan claimed.

I had heard that Rob ordered new farming gear for everyone: metal cages with feet that let the structures sit on the mud without sinking down. It wasn't floating gear, which everyone wanted except Dom, but it was a start. I asked Dan whether he had had any luck with the new stuff.

"It's sitting in my front yard. I haven't had any time," Dan said. "The Christmas season has been real busy."

Danny was sitting on a cooler, his arms resting on his belly. He fell asleep. Nothing roused him, not the policeman's whistle directing traffic, not the horn honks, not the motor hum. Joe pulled up in his red Tundra. I poked Danny in the shoulder, and with a grunt and a grumble he stirred back to life. Half awake, he climbed into the backseat and settled in for a snooze. Joe and Dan packed the truck bed and hopped in the cab. I shook Joe's thick-fingered hand and waved at Dan.

This good-bye felt awkward. Like mermaids marooned on a beachhead, the guys lost strength as they roamed from the coast. It made meeting them in this environment, bereft of their raw abilities except their wit, a strange occurrence. They had come to Atlanta, I thought, to promote native wild oysters among an unappreciative crowd, one that struggled with knowing the state's geographic boundaries. They left limping. The guys looked flayed by the affair, bushed, running on fumes. They had suffered the quips of humorless powerbrokers and enjoyed little of it.

Did the appearance help their cause? I wasn't convinced. What good did it bring? Certainly, Dom and the DNR weren't hurt by the display, and Dom's happiness would trickle down. For the watermen, Coastal Day had become an expected bit of peonage. They were, essentially, gratis caterers. There wasn't time or energy for lobbying on their part. But what if Justin had been present? What if he'd linebackered a few suits into a corner and explained the benefits of supporting aquaculture? Perhaps his resonant bass would have carried over the clang of serving spoons against hotel pans and all that political chitchat. It wasn't likely. Nothing could overpower the political vortex.

Unsure what else to say, I bid them home. Get back, friends, to she who keeps you in her favor. Leave the city behind. Stay faithful to the sea.

CLACKERS

Rafe Rivers in the Mud River.

In only a few short years, Rafe Rivers had become the premier organic vegetable grower on the coast. The best restaurants coveted his turnips and arugula. Shoppers at the Savannah farmers' market queued up for heirloom eggplants. By 2016, the Canewater Farm footprint had tripled. More rows, more crops, and more employees to bring it all to the table. Oysters fell by the way.

Justin felt as if Rafe, once the golden boy for the future of Georgia shellfish, had lost interest in oysters. He wasn't taking it seriously. Rafe hadn't been as hands-on with his oysters as Justin or Rob would have liked. At best, Rafe visited his oyster lease once a week, more likely twice a month. That wasn't good enough to keep oysters clean, healthy, and alive, Justin argued.

With a year-round growing season, Rafe's schedule was atypical for an oysterman. He couldn't opt to crab or shrimp in the summer. Now that he had revved up the farm and established markets, and dependable buyers, it wasn't something he could turn off. Oysters were a passion project that would pay off in the future, he hoped. Vegetables promised profits today.

Assuming Rafe hadn't lost interest, it seemed improbable he could ever find the time between planting, harvests, and all the other obstacles he had encountered.

Despite Rafe's success, the past year hadn't been easy. There were personal developments, like the birth of his first child and the unexpected death of his mother in January. Just as headstrong as Justin, Rafe persevered privately through these hurdles. Outside forces were another problem. The Department of Agriculture gave him a hard time when he attempted to set up the required wash stations and walk-in cooler for his oysters. His particular agent hadn't been forthcoming with directions, Rafe said. So, like a birthday boy whacking at a piñata, Rafe made mistake after mistake. He did everything wrong until all that was left to do was something right. His oysters grew fat on his lease, but without approval from the ag department, all they could do was grow fat. He couldn't sell them. Finally, in the spring of 2016, the state approved him. But the trial had dampened part of the fervor he had exuded when I first met him.

When I visited Rafe in March 2016, his frustration was evident. He wasn't angry—that wasn't his style—but he wore exhaustion like a wetsuit. Before, he would have camouflaged such minor weaknesses with hospitality and self-deprecating jokes.

"I'm not a quote-unquote waterman, and I do wonder what they think of me," Rafe told me as we walked his land. He talked regularly with Mike and Jeff. Often, all they had to say about oyster farming was, "This is bullshit." But the inspiration of Justin's example kept Rafe going, he said. Justin's positivity gave him hope. He worried about letting Justin down by not lavishing care on his oysters, as he had promised to do. He didn't want to fail.

"I'm not sure I should go full-on with this," he said. "Not until this all works so that I can push out maybe like a half million a year or something. I don't see the use in sending my guys out to harvest bushels. It's not going to be worth the money. I really want a farmed half-shell oyster. The chefs are so excited about it. I've been talking it up, getting them worked up about it, so I don't want to give up. I don't want to be the guy who didn't deliver."

In March, I traveled with him to his lease to clean his oyster bags of muck, cull dead oysters from the bags, and drop off two new cages, which Rob had given all the watermen. The cages had metal feet that would keep the oysters, bagged up within the cage, from getting lost in the mud. The cages weren't different from OysterGro cages, the industry standard in the United States; they lacked only plastic floats.

Setting up the cages would be easy. Cleaning the bags—retrieving them from the mud at low tide, culling them on boat, zip-tying them back on the farm, Rafe and I slogging through mud up to our thighs—would take hours. Our sunburns would sizzle for days, our legs rubbery and weak.

On this Mud River mudflat, Justin and Rafe had set up a system that helped oysters grow well. But for the farmer, it was a pain in his ass. Nothing, Rafe had found, was easy or quick about oyster farming on his lease.

Rafe and I culled bivalves on the bow of his boat. The dead ones were easy to spot. Their agape shells gave them away. Industry people called these *gapers*. Often, in supermarkets, allegedly fresh shellfish expressed this permanent yawn that signaled death, unbeknownst to shoppers. Watermen use the term *clackers*. Rapped against the hull or a shucking knife, the shell rattled like castanets. We tossed them into the river.

Mud collected on the bow, making a gravy-like slurry that soaked our clothes and gloves. We talked about how best he could farm oysters, how his operation might be improved. I had seen enough farms at this point—in Alabama, New England, and Canada, as well as countless pictures online—to see how antiquated this method was. He needed floating gear. It would make his life easier, bring him into the twenty-first century. I told him about what I had seen in Alabama, where oyster farmers used the Australian longline system. We talked about the floating cages employed by oystermen up north, especially in Canada. Rafe had seen similar systems during West Coast travels.

I knew of at least one oystering operation just up the coast in South Carolina that had begun using floating cages. It was a small operation, not unlike Rafe's or any of the guys'. This South Carolina oysterman was able to work at high tide, which made the job a smidge less frustrating. Unfortunately, I told Rafe, he wouldn't be able to use anything similar around here any time soon.

"I sure hate that we're trying to reinvent the wheel here, with what's been done so well in so many other places," he said. The wheel mentioned here being oyster farming, that floating cages were not yet allowed by the DNR. "It's like me going out farming with a plow and mule."

There were reasons that the adoption of oystering apparatus moved glacially. Up at Skidaway, Rob had been testing gear; he had come up with a design of his own he thought might work in Georgia marshes, and it was to Dom's liking. It used tumbler cages (a triangular hard-plastic case far lighter than larger floating cages) strung on ropes and marked by buoys. It wasn't too different from a crab pot, which was already permitted by the state. Dom was waiting on the data from Rob's experiments before he made any moves to introduce gear into the marsh. Permitting any kind of navigational hazard took months, Dom told me, and he wanted to be sure that the gear worked in Georgia and that watermen definitely wanted to farm with it. Dom also worried about homeowners and fishermen launching "not in my backyard" complaints. He had already heard a few whines reflecting some of the "interesting socio-dynamics" that were bound to flare up between shellfish farmers and the marsh's other stakeholders (naturalists, conservationists, recreational fishermen, people who wanted the views from their marsh-side property unobstructed by aquaculture). A battle between blue and white collars might be fought. His worries were far from unfounded. On both the West and East Coasts, community uproar and lawsuits were common when it came to where oyster farms were allowed to operate—not in my wetlands, they cried.

"Shellfish [aquaculture] is great," Dom had told me. "But if it results in what's considered an eyesore to some, there's a problem."

Dom had heard envious quibbles from a few oystermen, who asked, "Why can't we have what they have in South Carolina?" Because there were questions Dom needed answered: Does the gear really work in Georgia? Are watermen truly willing to invest in the gear? Are homeowners going to pitch a fit? Dom waited for data from Skidaway before deciding. If the numbers looked good, and the industry wanted floating gear, he would begin the process to approve it. Until then, Rafe would have to make do.

Rafe and I cleaned and culled until it was almost dark. Spent, we headed back to land. I was convinced that Rafe wasn't going to give up; he was too proud for that. But given the list of projects he kept in balance, he would need support. Not free labor, although that wouldn't hurt. Just a relaxing of rules and a bit more guidance from regulating bodies.

"I just wish there had been a class for this," Rafe said.

RUNNING THE RABBIT

Justin and Amelia Manley.

Throughout the spring, Justin and I tended our relationship by phone. Those conversations felt tense. Justin sounded stressed, even worried. I chalked up the anxiety I heard crackle through the cell receiver to be typical workplace drama. Justin was headstrong, if nothing else, and it was clear that working within a hierarchy as rigid as the university system wasn't the most comfortable fit. What transmitted through the phone were the last growing pains in his maturation from waterman to lab man.

Justin would be fine. There was no real need to worry about someone as resilient as he. Yet I wanted to better understand the terms of his transition, to find out what deal he had struck with himself and with Amelia.

Over three years, the myths and expectations of the Spatking had fully collapsed. But they had been replaced by something honest. In the process, flaws had been exposed. Justin mended them slowly.

In June 2016, I drove down to Skidaway to attend a GSGA meeting organized by Tom and Justin to catch growers up on the status and future of the hatchery. The evening before the gathering, Justin and Amelia had invited me to dine with them at their home one more time. I steered my car from the Harry S. Truman Parkway onto the spur toward Skidaway, a maneuver I'd made more than a few times—not as frequently as the locals, but on enough occasions to have flipped the right blinker with confidence. All these commutes had become comfortably routine: to Skidaway, to Earnest's, to Joe's. I hoped they would always remain fresh in my mental map, with Justin's coordinates bolded on the key. But almost as often as I had pulled into the Manleys' gravel drive, I had almost missed it. Neighbors' bushes and sloping tree limbs hid the house's yellow paint—always a reminding detail for me—like a curtain. The endless oak wall that hugged the rumble strip offered no sneak peeks. Only at the last moment did catching quick sight of Amelia's SUV—the bone-white one with the OYSTERS vanity plate—or Justin's pearl-colored Detroit chariot prevent a reroute.

Justin had served as my first guide along the coast; he taught me which channels, either tide-based or paved, would take me where I needed to go. Although I now recognized enough landmarks to lead me from Harris Neck to Sunbury to Skidaway without detour, if I was to make it in time for dinner—to even locate which house contained the correct dinner table—I still relied on the Manleys' vehicles like cardinal beacons. At the final nick in the road, trunk and tailgate alerted me to peel into their driveway, a shovel of gravel kicked up as wake.

The Manleys' place relaxed the mind with stunning visuals. The sliding-glass back door transmitted light at full spectrum, the palette of the marsh brushed across white walls, hues of salmon, marigold, and lime.

The Manleys had only recently returned from a vacation in the Cayman Islands. Their skin still smoldered from the sun—their faces arid and red from ultraviolet exposure and salt water and a few too many Caribbean lagers. They wore shorts and tank tops, the grill flared on the back porch, and, enviably, their clocks were still set to island time.

From the Caymans, their children had traveled to Toronto with Amelia's parents. For the next few days, they were kid-free. On Friday, Amelia would drive alone up north to fetch them while Justin stayed behind to launch another spawn. This was a loose evening before settling back into the routines of work and home.

I watched Amelia prep dinner and asked whether she still thought about moving. She nodded. She wanted to go back to Canada.

What about you, Justin? Did the resentment he expressed in the spring remain? Did he want to leave?

It depends, he said, on what happened in the next year. Would there be a job after the temporary funding lapsed? As he improved as a hatchery manager, was there a viable industry to grow his oysters to market size? On both counts, he hoped so, but those weren't questions he could answer himself. For a man used to wrestling obstacles into submission, his powerlessness made him restless. He felt that he was doing his part to push the industry forward, but he had worked so hard that it all seemed lopsided now; the industry side had to catch up. As Joe and the others got older, the window narrowed, he worried. On the growers' end, someone had to pick up the baton and run, he said. Justin, on the sidelines, could only wait, ready to click the stopwatch.

What about the politics of it all, the university bureaucracy? Did he still feel like a caged animal?

He still wasn't allowed to talk to anyone, which didn't make any sense to him, but he had moved on.

"I remove myself completely from it and focus on what I'm doing," he said. "I make my own personal impact, and that's all I can ask of myself. I put my head down and do what I do well."

In scientific professionalism, in the practice of husbandry, in mating egg and sperm for spat, he found peace. He knew he was needed: if he walked away and took his expertise with him, he wouldn't be easy to replace. The lab would be left with a deficit, Justin believed. But in the end, he didn't want to run. He had been enjoying the lab lately. He and Rob and Tom were friends, and with the interns and students around, the lab now reminded him of when he was in grad school, when a dozen people worked at the lab, studying or researching, when Randy Walker had been in charge. That memory made him respect not just his corporeal connection with oysters, but also with Skidaway, with Savannah, with Georgia.

When he considered how much time he had given this particular animal, this particular stretch of brackish water, to walk away would look sophomoric. He couldn't quit, not now. He had made an investment in Georgia oysters. It would be stupid to throw away the equity he had sweated for the industry. Maybe they were still a ways from stability, but they were close. If the legislature didn't come around, or the hatchery work began to feel futile, maybe he would consider a move. He didn't want to, though, since the opportunity to make Georgia oysters famous still inspired him.

We strolled off the deck and into their backyard. In the past year, the Manleys had constructed a handsome fence that served as a barrier between their children and some raucous neighbors. A brick path led down to an outdoor patio, where we had gnawed our way through grilled chicken. We walked on, toes in Bermuda grass, skirting the raised-bed garden where Amelia had been growing tomatoes and where Justin raised hot peppers.

We continued onto a wooden walkway anchored into the last few feet of the Manleys' lot. As the platform extended into the marsh, the earth angled away from the treads; the turf grass became spartina, then thinned out into patches of mud. We walked onto the gangway toward a gazebo stationed thirty yards out into the marsh, among tidal creeks that leached northward into the Vernon River. Halfway across, the gangway crooked and bowed under our weight. We lurched left and right with each unbalanced step. An elevated tide from a recent storm had unmoored the framing from the piling that kept the walkway stable. It still worked, and with caution, we could traverse the route out to the overlook.

From the platform, Justin pointed out clumps of shells plastered into the creek bottom. They were the remnants of old experiments, he said, nonviable animals whose bodies needed a final resting place. If he dumped enough slag like that, he joked, maybe he would eventually have his own oyster mound.

Across a crew cut of cordgrass, the living-room lights of the Vernonburg community began to twinkle in the dusk and shine back at us. If I let my vision blur, and let the thrum and blink of traffic on the Truman Parkway dissolve as if in a hypnagogic state, the stars on this summer night might glow like the guiding lights of antique bateaux, sloops laden with an evening's oyster harvest.

The sun sent up flares of retreat across the wetland scalp. It was a scene that could calm any restless heart. Although tides washed sediment along the river bottom, adding to and replacing the layer twice a day, the marsh suggested the same tranquility of permanence as a mountain range. I believed that peace to be a balm for a man as exertive as Justin.

Justin's story wasn't finished; the rebirth of his beloved Georgia oyster was just as incomplete. He would continue to push for a fishery he thought should adapt; that fishery had slowed his momentum to a speed suitable to its nature. I expected the two forces to clash for some time, perhaps finding a conclusion long after my interest in the battle had faded. The only closure I could find in this journey was a final freeze-frame of this coast and culture, a last impression from a discrete moment in a protracted history. I hoped to distinguish this milestone by exchanging a sort of farewell in the gloaming.

"This place is so beautiful," I told Justin and Amelia, unable to draw out anything more poignant. Their faces radiated orange in the twilight. This feeling had to be related to the peace of mind that Earnest McIntosh found out on the water. "I can't believe you get to live here."

"I know," Justin said. "We're pretty lucky."

When I complimented the natural grace that greeted us this night, that closed many evenings at the Manley residence, I hoped to honor the landscape that Justin had first showed me to be worthy of admiration for the bounty it produced. Without him, I wouldn't have encountered the fascinating people who worked the marshes, or the deep friendships and idiosyncrasies that animate this evanescent wild. There was more: I wanted that fumbling gesture to eclipse flattery—the statement was personal. It was an ovation for a life I admired. Maybe I botched it, played it cautious and missed, but I think he got the point.

The next morning, the shellfish lab teemed with a life in a way unseen before. Two years earlier, Justin had introduced me to a research office that specialized in boxes and cobwebs. He and Tom, and later Rob, shared turns cranking the lab's battery back to full charge—cleaning, scrubbing, building, spawning—and it took some time. Almost a year after I first watched Justin mate oysters here, the physical difference that their hard work had generated was impossible to miss. Justin's hatchery had outgrown the original wet lab. A barnlike structure to the rear of the lab, white and two stories tall, was the only place to expand. Used in clam research before Justin's Skidaway tenure reset, the barn had been used as storage for all manner of things: nets, shellfish bags, boots, wetsuits, lumber. Wide windows on the upper level let light stream in, and removable walls on the side that faced the lab allowed large equipment to be slid inside. After a bit of spring-cleaning, the lab crew installed three upwellers in the barn and used a spare room to store oyster brood stock in a wet tank between spawns.

"We used to have a hard time keeping oysters alive," Justin told me. "Now we don't have enough room for them."

As far as problems went, the latter wasn't the worst to have.

Youth accounted for part of this summertime transformation, at least externally. A high school intern and a graduate student leapt up the staircases between the lab's two stories and scurried between the barn and the dock where the lab's small research vessel stayed tied up. They were quiet workers, but their swift movements gave an energy to the place.

Rob Hein, in charge of the student workers, and still quite young himself, led the buoyant bolt the lab had received. On this humid June Tuesday morning, I found him bedecked in board shorts that parachuted past the knee, a long-sleeved T-shirt, and a soaked wet bandana wrapped around his face like a balaclava. The sun would be brutal, and Rob was prepared. The semitropical heat lamp wouldn't stop him: he stomped around the

Rob Hein working oysters in the Skidaway River.

grounds outside the lab with determination, like a foreman unafraid to swing a hammer himself.

On a cement pad between the lab and the storage barn, Rob and his helpers had set up a row of polyethylene tubs, five in all, the same ones used as downwellers in the hatchery. They came up to Rob's thigh. In recent months, algae—the stuff oysters dine on—had piqued Rob's interest, and he had made growing the lab's own algae for oyster larvae a personal project. Until recently, new oysters had been fed from algae ordered online and kept in a fridge until a spoonful or two was needed. Justin, Tom, and Rob thought it better if their larvae fed on baby food that more closely resembled what they might eat in the wild. I was in time to watch Rob prepare a meal; his kitchen consisted of those five blue containers.

Rob jumped in and out of the tubs. He checked and replaced filters on water pipes that ran into the tubs through circular holes cut into the sides, then hopped back out. Soon, river water came rushing in, filtered so that only algae, bacteria, and water penetrated the screen. Once the tubs were filled, the water cycled between them through pipes and foamed and bubbled. Rob added a small dose of nitrogen to the water, just enough to force an algae bloom. In the ocean, an algae bloom was something to fear. Rob used a bit of Miracle-Gro, the type of fertilizer that Charlie Phillips wished upriver homeowners would keep off their lawns. But within the hatchery microcosm, the bloom, caused by an over-the-counter fertilizer, made a buffet that developing oysters could gorge on. In time, the water in the tubs turned brown and opaque, a good sign, Rob said.

While I pursued Rob's outdoor algae efforts, Justin was inside the lab, blowing air onto downwellers to dry them out before setting eyed larvae. Bacteria populations thrived in warmer months, and their presence could plague his June and July spawns, Justin told me. So he cleaned and cleaned again. During a break, he joined us outside, leaning over the algae tanks.

"We're going to need an algologist for long-term success, don't you think?" Justin said.

"For sure," Rob answered.

The idea of hiring yet another person at the lab perplexed me. Was there money for such a specialized job description, a staff member devoted to growing algae?

Here is what I knew had occurred in the spring: the legislative session ended with no mention of funding the shellfish hatchery. In the reconciliation process that sent the budget between House and Senate, its line item had been erased. The document sent to Governor Deal for his signature offered no money to the fledgling oyster-farming industry. When I first heard about the deletion, I emailed Tom Bliss to ask what he intended to do. He had no funds lined up, but he was looking at his options, he wrote.

During a phone call with Dom Guadagnoli over the summer, he underscored the importance of support from the Gold Dome. "If you look at what's happening in Alabama or in Virginia, and in the whole Delmarva Peninsula, the progress that's been made [in oyster farming] there, it was all in conjunction with a legislature that was in full support. It was not grassroots."

Dom wasn't a lobbyist, he reiterated to me, it wasn't in his job description. I was to infer from his comment that if someone leaned on the flesh pressers in Atlanta, it wouldn't hurt the cause that Dom, Tom, and Justin were working hard for. If the industry was to grow, it would only happen with political aid.

Despite the erasure of the $150,000 request from the budget, Tom labored on. As the algae-bloomed water began to flow over the barn's three upwellers, Tom joined us outside, carrying his coffee mug. His long hair was pulled in a ponytail, as usual, and thin-framed glasses sat on the bridge of his nose.

I asked him what he knew about what had happened with the legislature.

"Politicking," he said. "Something behind closed doors."

He said that even the president of the University of Georgia, Jere Morehead, had no clue. It happens, Tom said, and they weren't alone. Every cycle, some project became a victim. There was no aggressor, nothing to blame but whatever political wind blew that day.

"This would've been a big bang for their buck," Justin chimed in. "It's not that much money, there's existing infrastructure, and there would be an impact."

"I've been told that we didn't ask for enough," Tom added. Cutting $150,000 was easier, I guessed, than $500,000 or more. "I've been told we have to not be so conservative. It doesn't make sense, not to me," Tom said.

Luckily, Tom had had some success in finding alternative, short-term funding for the lab. With the help of President Morehead and a vice president of outreach for the university, a small cache of money was allocated to the lab to keep Justin and Rob employed through the summer of 2017. Tom had been given a window to expand grant-writing efforts, and he and Mark Risse could mount another attempt to secure a spot on the state budget.

"Morehead has been down here, and he likes what we're doing. He believes in this," Tom said. Perhaps they had found a lobbyist in one of the most powerful men in Georgia. Time would tell, since the spring legislative session was still months away.

"In time, the hope is we become self-sufficient by selling seed at cost," Tom said. "But there is a couple-year gap until we get there."

State funding would help keep Rob and Justin employed, Tom said, and keep the lab afloat until that gap was crossed.

Did he feel confident about the legislature?

"Yeah," he said. "I do."

Dom, in our phone conversation, also felt "hopeful" that the spring session would bear fruit.

I had come to Skidaway this June to sit in on a gathering between GSGA watermen, the shellfish lab, and Dom, representing his official capacity at the DNR. It was to take place that afternoon. At the meeting, Tom planned to inform shellfishermen about the amounts and size of the seed they could expect to receive in the fall, and to offer them a little transparency about what was happening at the lab. A number of the watermen I had spent time with over the past two years would be in attendance, giving me an opportunity to catch up with them all at once.

Justin seemed happy. I'm sure the knowledge that he would be safely cashing paychecks for another year gave him peace. Certainly, part of his good cheer stemmed from me finding him in the middle of his busy breeding season. In gearing up for the Friday spawn, which would require near round-the-clock attention on his part, Justin's workload reached capacity. This was how he liked it. Just tasks, boxes to check, no idle time to waste. The procedures gave him unquestionable purpose.

Justin looked forward to seeing the GSGA crew again, even as his connections to them continued to loosen, given Rob's devotion to fieldwork. Justin had made peace with his diminished public role, but still enjoyed time spent in the guys' company, no matter how fleeting.

I asked who planned to make it to the meeting. Rob had heard confirmation from Dan DeGuire, Joe Maley, John Pelli, Earnest McIntosh, likely Charlie Phillips. It was a no from Jeff Erickson, and likely Mike Townsend, too.

Rafe wouldn't make it to the June meeting. His wife was sick, and his mother-in-law wasn't able to help care for their daughter, so Rafe stayed behind. The rest of the expected crew made it: Charlie, Joe, Dan, Earnest (with a grandson, a recent high school graduate from Odessa, Texas, tagging along), John, and Dom. Tom's boss, Mark Risse, drove in from out of town.

We gathered in the same classroom where the first GSGA meeting I attended took place. Two circular tables were set up with stackable chairs ringing them. Tom had set a stack of printouts on the table, copies of a rough agenda he had written.

Everyone arrived roughly on time, and Tom kicked the meeting off nonchalantly.

Brood-stock oysters.

"Mostly we want to fill y'all in about where we are with the hatchery," Tom said. "We've had two successful spawns this year, all the way to set."

"Estimates put us as producing between 2 and 2.5 million seed," Justin added.

"We're going to keep spawning, keep working on our skills, and practice the husbandry, the timing of it all," Tom said. He explained the stopgap funds allocated by UGA to keep Justin and Rob working. "We're stable for now."

Two grants that Tom had in the pipeline sought to explore new marketing techniques for the burgeoning industry. Since GSGA watermen were too small for distributors to deal with, perhaps shipping via FedEx or UPS posed a viable route to potential clients in Atlanta. He was also looking into lease mapping, which would give both the lab and the watermen a more thorough picture of what was possible in their corners of the marsh.

Tom told the guys that the lab was stepping up its site visits. From now on, Rob, Tom, or Justin would be visiting each waterman at least once a month. That way, Tom said, we can assess what is happening, what your needs are, what the oysters' needs are.

"We've got a ton of data that we need to get out to y'all," Tom said. "We need to know what locations aren't working for you and change gear depending on your lease."

"If there's a spot on your lease you think will work, let us know, and we can look at it and think about the right approach," Rob said. "We can come up with summer and winter plans."

Tom asked if the watermen had any feedback on their oysters.

John Pelli said he'd seen some mortality, about twenty to thirty per bag. Earnest said his were doing good, living longer, and growing really good. Dan saw a high survival rate and very little fouling. Charlie said he hadn't checked on his in a while, but some yacht wake had blown over his bags recently.

"Those cages you gave us, they threw my back out for a week," Earnest said.

"Yeah, for small farmers like you guys, the tumblers that Rob had been working with are going to be best," Justin said.

Tom turned the meeting over to Dom, who wanted to gauge opinion in the room on a particular topic. Two or three phone calls a week still came through his office, asking for a lease. The waiting list was unbelievable, Dom said, "a myriad of people" that filled nine pages. Fifteen years ago, he joked, we couldn't give away a lease, and now this.

Dom's questions: Are we ready to open the industry up to new lease-holders? We have a lot of unresolved information that needs gathering. Do we need more members to help answer those questions?

"We want to know that the people who come to work a limited site can hit the ground running with capital investment," Dom said. "We have a lot of contenders."

Charlie spoke first, to urge another to respond: "Come on, Joe. I've already talked his ear off about this." I had heard that Charlie had been calling private meetings with the other watermen—UGA wasn't invited. Charlie used the closed circle, I was told, to further complain about state regulations. With fewer rules about importing seed and gear, as he had told me in interviews, Charlie could really make some money. Charlie lobbied to lock down new leases, I presumed, to dampen any competition he might face down the road. "We've got to decide how we're going to run the rabbit."

Joe leaned toward keeping the industry closed, for now. "We are still in the baby-step stage," he said. "And I don't know how much more capacity Justin can support."

"We think we might be able to push out ten to fifteen million in the current setup," Tom said.

"Is there seed capacity for more people?" Dom asked.

Tom and Justin thought so.

None of the other watermen spoke up, giving the impression that Joe spoke for the lot. Mark Risse, a tall gray-haired man, with a voice that conveyed power in its low timbre, disagreed with the premise that an unstable seed supply precluded introducing new watermen.

"It's state land," Mark said. "It's my opinion that the state should be letting qualified people have access to a lease. It's not the university's job to have seed supply in place before opening the gates. What industry does that?"

An argument could be made for keeping the industry closed until the grant period ended and the watermen collected all the free seed promised to them, but it wasn't proffered.

Later, Justin whispered something in my ear, echoing the same sentiments we had shared many times over: "The old guard doesn't want the competition, but I think the state knows that it can't stay closed forever. There has to be some new energy in here."

Dom's biggest fear was that if he unlocked the industry sooner than later, a large aquaculture conglomerate might gobble up a lease and push out small-time growers. A healthy fishery, he contended, featured many people of varying economic backgrounds. The day's debate didn't find a conclusion, but it ended as it began, with Dom confirming that, yes, the watermen wanted the industry closed, at least for now. It had been vocalized in murmurs during the meeting, but it was obvious that private discussions had recorded more-fervent opinions.

"Well, we hold our breath," Charlie said, hoping to close the meeting with a final comment. "We've been doing pretty good, but we've got a lot of work to do. I want people in. I just don't want them running into a brick wall."

Tom offered to give the guys a tour of the expanded hatchery, to explain algae feeding, and to show them the new crop of oysters. I stayed behind to catch up with Justin, to have him tell me which lines I needed to read between.

Dom's question wasn't really about leases, Justin told me. It was about seed. Dom alone could lift the ban on imports, and he was evaluating whether he should. Charlie pressured him regularly on the matter, Justin said, and Dom had begun asking UGA how it felt about the matter.

Well, I asked, how do you feel about it? You were such an advocate for the ban. Have you changed your mind?

Upon consideration, Justin now felt it safe to import seed, but only from certain locations. Georgia's clammers had been buying seed from a hatchery in Tampa for ten years. Any disease that might have come in from that hatchery had already established itself, and we are dealing with it, Justin said. Since that hatchery also produces oysters, why not let some in? It wasn't a bad idea. He still felt imports from New England or Virginia were dangerous. Too much risk. People think that because all eastern oysters are the same species, they should be interchangeable between habitats along the Atlantic Coast. But oysters developed traits based on locations. There was much we didn't know about diseases; there could be other parasites that flourish in warmer waters. More research was needed.

Anyway, he said, there wouldn't be enough seed available to solely supply a Georgia farm, but there would be enough to help the guys out, make them a little extra money and get them to market faster. Dom's plan, Justin said, might be to allow minimum seed import, a small number, just a bump to placate Charlie and aid a few others.

As if he had become a politician, his views on a contentious subject had evolved. No doubt a result of his pilgrimage from marsh to laboratory. Quite a shift from where you were only a year ago, I told Justin. He understood what I was getting at. The change of heart on seed made sense, he said.

We caught up with the tour, and Justin and Rob gave an algae lecture to the guys. A few oystermen, like Dan, soon departed to beat commuter traffic out of Savannah. The stragglers retreated to the barn and the upwellers. They circled around Dom, who held his notebooks to his chest and listened to an enumeration of the hopes and dreams of watermen and scientists. The water pump buzzed loudly, and the surge of algae-filled water flowing into the upwellers made it hard to hear anyone more than five feet away. No matter whom we spoke to, we strained to catch syllables bouncing off stray droplets.

I chatted with Earnest about his grandson's upcoming move to college. Quite the climate shock to move from West Texas to North Dakota, I said. Earnest paused to bow toward me, point at Dom, who commanded the gaze of the group nearby us, and say: "That's the man right there. He's got all the power."

Watermen were lucky to have a regulator like Dom looking out for them. He was a stickler for hewing to the rules, but he knew it begat longevity. Dom considered it smarter to seek permission rather than beg for forgiveness. Ask all the questions now, exhaust every inquiry, then proceed with perfection, no need for apologies. He knew it pained watermen to move so slowly, but everything had to be just right. As the enforcing agent in control of their livelihoods, Dom was afforded a parental respect by the watermen—upsetting or disappointing him was the last thing they wanted. In reality, Dom had a boss and followed directions, too. In truth, he needn't have been feared. Dom strove to build consensus and partnerships; that was the sense I got from him. This wasn't a marshland autocracy—Dom was an advocate. He set rules while trying to please everyone. He was a friend, yes, but for watermen and for UGA, the implementation of new ideas started and stopped with Dom. And for that, he commanded everyone's attention.

Tom gushed to Dom about the potential projects the lab could launch. Oysters and clams weren't the only shellfish that UGA could research and breed. Mussels, blood arks, cockles—"There's a plethora of shellfish to go after," Tom said. "[Much of the] Gulf doesn't have the diversity we have."

"I'm super impressed at the progress y'all made," Dom said. "You've come a long way."

It was an approving comment that Tom, Rob, and Justin were relieved to hear.

"Honestly, it was life or death," Justin said. "We had to succeed."

Mark laughed: "Not like it was your job or anything!"

OYSTER SOUTH

A hatchery-raised oyster.

On a crisp evening in January 2017, Justin and Tom arrived at Acre, a restaurant in downtown Auburn, Alabama, as valet-driven cars zipped in and out of a tight parking lot. Outside, finely dressed diners waited for tables. Men stuffed hands into khakis or wrapped arms around dates to provide warmth.

Inside, gathered around the restaurant's bar, dozens of oystermen and marine scientists picked at hors d'oeuvres. In town for the inaugural Oyster South conference, hosted by Auburn University and organized by Bill Walton and Bryan Rackley, they mingled within an interior that mimicked the decor of a hunting lodge. Rough-hewn trim wood dampened lightbulbs that flickered like candle wicks through faceted glass shades. Food presented buffet-style on thick slabs of felled and milled tree trunks looked equally primitive. Deer and boar charcuterie and bowls of sour, sappy sauces gave the impression that wildlife had been invited indoors. Some of the guests were far wilder than the food.

Watermen from Apalachicola and the Outer Banks spent the day learning about the status of fisheries along the South Atlantic and Gulf Coasts, reported by professors and extension agents from several states. They talked marketing strategy, for example, how to best attract and keep restaurants as clients. They exchanged business cards and spoke of successes and failures.

At Acre, they after-partied. Free wine and whiskey samples encouraged looser conversation. Some watermen looked more comfortable than others to be wearing collared shirts and talking close under romantic lighting. They were hundreds of miles away from the bays and rivers they worked daily. Acre's rough interior conjured an untamed past before the present's digital intrusions, when hunting and heirloom seeds were quotidian rather than fashionable, but it was likely nothing like the boats and abodes they had left behind.

Watermen came from rustic digs where authenticity was measured by practicality over good design. Yet these two dissimilar worlds connected with increasing regularity. Acre, like the Kimball House in Georgia, was the kind of place where raw oysters sold quickly. Strong partnerships between water and kitchen were necessary to push aquaculture forward. Even watermen who might never afford an entree in such an establishment required fluidity between the classes. They adapted to the surroundings for survival.

Justin looked uneasy here for different reasons. He had stuffed as much knowledge into his cranium as possible during the day, just as everyone else had attempted, but somehow his concentration seemed more studious than the others'. He hardly spoke and rarely averted his eyes from the projected PowerPoint presentations and podium speakers so as not to miss a word. His bathroom breaks were few. He never rose to refill a coffee mid-session. He took this continuing education opportunity seriously, and it gave his demeanor a sternness that proved difficult to soften. Nervousness caused this.

At Acre, he was surrounded by oyster dignitaries, scientists such as John Supan from Louisiana State University and Bob Rheault of the East Coast Shellfish Growers Association, and closer to them than at the conference. The same experts whom he had absorbed that morning with intensity now strolled through the restaurant, brushing shoulders with him. The immediacy of it all made Justin, who had become far less verbose as his extension career lengthened, more mum than usual.

The experts nodded, shook hands, and exchanged introductions and pleasantries with Justin, clueless about how shy he felt in their vicinity. To Justin, they were a community he had admired from afar but had not yet joined because he didn't feel prepared. To describe these career aquaculturists, he used the word *family*, even though they didn't know his name. That was how finely he followed their research: he memorized published papers like branches of his genealogy.

He froze in their presence. He passed up free drinks to keep his mind focused on the opportunity, hoping to glean whatever new knowledge he could. He didn't want to embarrass himself, and he definitely didn't want to miss some nugget of insight.

These scientists knew what Justin was going through. They had been there themselves, only years earlier. They understood the struggles to hatch oysters and keep them alive, and the often awkward interactions with fishermen. At the conference and during this soiree at Acre, Justin hoped to glean this wisdom. His shyness among this crowd showed reverence. His reticence helped hide anxiousness.

In a corner behind a booth, en route to the restrooms, oystermen from Alabama and North Carolina opened coolers filled with ice and oysters. The watermen shucked and presented their wares to eaters, and inspected reactions as the shells were upturned into mouths. Lane Zirlott, mouthpiece for Murder Point, a Bayou La Batre, Alabama, shrimp company that had redirected itself toward oysters as other catches faltered, enticed people with appeals loud enough to carry over the crowd. Dressed in a patterned shirt with a tall collar, his hair coiffed into a spiky Mohawk, Lane was as fashionable as a celebrity chef. He smoothly shed the salt-flecked uniform of an oysterman for club wear. In turn, his sunburnt skin could be mistaken for a vacation memento and not the base layer of a workingman. In turn, few oyster companies had ingratiated themselves onto menus around the South as well as Murder Point. Part of that success was attributed to the product. The rest was Lane's presence. He could grease business channels with a slick and vivacious Mobile Bay patois that made deal making a foregone conclusion.

He pried open oysters that he drew out of an insulated chest and handed them directly to eaters, marrying the drawl of the blue-collar bayou to fine dining's deluxe britches. It was a welcome and entertaining show.

After downing nearly a dozen myself, and basking in Lane's spotlight, I drew Justin over to try the selections and gauge his reaction to interstate competition. I started with a Murder Point, handing him a shucked shell and joining in with him to sample. We both loved the dark shell and fat meat of Lane's oyster. Its plump pocket of mild brine was easy to chew. But it didn't fit our personal tastes.

Justin downed a Cape Hatteras easily. He reviewed the slim, chalky shell and looked pleased.

"This isn't that different than what we have," he said quietly. Crisp and salty, like spring wind sweeping across a barrier island beach, but not nearly as slashing and grassy as a Georgian, its liquor diluted compared to the bottomless salinity of a Harris Neck. I took Justin's comment to mean that he felt Georgia oysters deserved a place at Oyster South. I did as well. The product from Georgia was already quite good. Quality was in place, but it was all that other stuff—bureaucracy and personalities, the less tangible features—that barred them from the party.

An oyster from Ossabaw or St. Catherines would earn recognition one day, just not today. To this point, Justin offered a shrug. He was comfortable with the tardiness. This project advanced in its own manner, he said, with a bit of lament. It couldn't be rushed, for better or worse.

Justin, Tom, and I stood at a bar table and chatted, joined by Jay Styron, owner of Carolina Mariculture. That morning, Styron had shown a slide that detailed the distribution of shellfish aquaculture revenue on the East Coast, based on data from the East Coast Shellfish Growers Association.

Virginia, as expected, dominated the pie chart. Mostly because of its prominence in the clam industry, Florida was a distant third behind Connecticut. If Alabama's Gulf Coast had been included, the state would surely have taken a nice slice of the pie. Alabama was definitely on pace to gain more. Despite that state's tiny coastline, there was a reason Oyster South was held in Alabama and not Mississippi or Texas or Florida. Extension agents and entrepreneurs—the Zirlott family and Bill Walton's team—had worked swiftly down in Mobile Bay to earn their market share, and their efforts had become a beacon leading the rest of the South.

Among the thinner sections of the East Coast chart, North Carolina held a slim slice with 0.4 percent. South Carolina's was even smaller at 0.2 percent. Thankfully, they had given Georgia a serving. An invisible zero. Less than 1 percent, such insignificant landings that the exact amount didn't warrant inclusion. The number shocked no one. Few expected serious output from Georgia, regardless of Justin's and others' battle to buck the century-old tradition of wild harvest.

The chart was based on numbers from 2013, about the time when Justin and I met. The totals had assuredly shifted some since this four-year-old sorting. But the rankings hadn't. Georgia still scraped the bottom, while the Carolinas had leaped forward.

What Tom and Justin had accomplished made waves in McIntosh and Liberty Counties, but compared with the state's Atlantic and Gulf Coast neighbors, it would require a sprint to catch up to others' advancement.

I asked Styron what explained their success. Initiative, he said. Eight years ago, nobody was paying attention. The natural resources division in North Carolina didn't care about oysters or aquaculture. So Styron and other growers banded together and went straight to the legislature. They skipped over regulators, the Carolina equivalent of Dom Guadagnoli. The economic development argument was an easy one to make, Styron said, so they went to the statehouse to make it. Rule changes and funding soon followed.

Tom and Justin looked at each other and admitted that the GSGA could do something similar. They just needed someone to lead the charge. They wouldn't name names or blame any specific watermen for not championing the industry when it was within their power to do so.

"I guess I couldn't see Joe Maley up in Atlanta lobbying," I said. "Charlie is the right guy to do it, but he won't, right?"

Silence met my comment. Justin raised his eyebrows as Tom looked down at his glass, the ice cubes melting in a pour of bourbon.

I immediately thought of someone who could fill the advocate role. The industry needed the Spatking, now more than ever. Since Justin had retired the mantle, no one had stepped into the role as pitchman for the rebirth of the Georgia oyster. Tom and Justin couldn't accept the position; it wasn't the job of a scientist to do so. Their job wasn't to promote, only to research and assist. They were extension agents, not lobbyists.

A leader had to emerge naturally, someone, perhaps, with a personality like that of Lane Zirlott, who possessed a gabby manner and a willingness to pack a social calendar full of oyster promotion like no other. Without a leader, progress would always come slow. Justin would breed resilient oysters, but there was no promise that a grower would tend them successfully. Justin had come to terms with this reality. There was no other option. He couldn't hoist this burden over his shoulder and trudge through the mud, against the tide. His job was to watch and help, if asked.

If he still felt frustrated, he had learned to keep it quiet. As he, Tom, and I wondered about the possibility of following the path set by North Carolina, Justin stifled opinions in front of Tom and his colleagues. There was plenty to be said when Styron talked of going over the heads of regulators. More than anything, hesitation on the part of natural-resource agents to adapt marine laws to modern aquaculture slowed the growth of the Georgia oyster industry. It was a fine point that I could make and be frustrated by, but Justin and Tom knew who set the rules to the game they played. They hewed smartly, and politically, to the rules. It required great patience on their part.

Despite their shrewd composure, their faces leaked tells. I knew Justin's visual cues, so when he curled a cordgrass-thin half smile or looked over his glasses at me with the wing speed of a seaside sparrow, he clued me in to his thoughts. I had heard the complaints before. The signals recalled what he had said about the subjects in the past. There wasn't much point in giving them volume again. He had cached his objections among the periwinkles and humid inertia in the estuary.

The next afternoon, Justin and I met at his hotel on Auburn's outskirts. The oddness of the location seemed notable, about as weird as a bunch of watermen sipping California vintages in tony eateries. Back in 2013, our conversation began on the water. Over three years, our talks had continued in cinder-block offices, in wet labs, and around dinner tables. The water always remained in sight. Now, we met in a lobby nook on the Alabama plains. Except for the whizz of a nearby automatic door, the pre-check-in lull offered us uninterrupted quiet.

With a few hundred miles between Justin and the nearest salt marsh, perhaps he could offer a clear perspective on what being in the company of oyster luminaries did for his confidence in his abilities. Did being among oyster growers and industry advocates who, despite their own particular hindrances, could now share triumphs make Justin feel inadequate? Or was he renewed with inspiration? I hoped Justin would speak freely away from others.

"They used to call us shitty coon oysters," he said. "And when I came [to Georgia] in 2003, you couldn't even farm. There was zero oyster seed. We've made progress. From what we've come from, it's leaps and bounds."

I had seen a picture of the latest brood stock Justin had produced after another year of spawning and selecting. It was perfectly small, with deep frills along the edge and a strong gradient of gray tones. The specimen reminded me of how Justin had first described his vision of the perfect Georgia oyster. A petite shell with a deep cup and a wavy lip, a southern Kumamoto.

This picture offered proof of Justin's success at the hatchery. You did it, I told him. He sheepishly smiled.

"It took a village," he said. "We're doing a lot, with nothing given to us. That's our story."

For the work that he, Tom, and Rob conducted at Skidaway and with oystermen, the $250,000 grant they had used to get this far was paltry. The university supported their work in spirit, but hadn't been able to make any more cash materialize. "Nothing given to us" meant, for Justin, that only something closer to carte blanche could push the hatchery and the industry where they needed to be. If only they would invest in us now, Justin dreamed, it would be like buying stock in a company of computer nerds working out of a Los Altos garage in 1976. Hyperbole, sure, but he earnestly believed in the possibilities of this particular oyster and this particular band of oystermen. Not everything, of course, would work out.

Oystermen would quit trying to farm, citing their hobbyist interest in the trade. Some would retire. Some would give up leases in pursuit of a more manageable income stream. New oystermen would sign up. Dom would take aggressive steps to change regulations in aid of oystermen. Triploids would come. The state would permit floating cages. The industry gathered prerequisites for its renaissance slowly, and Justin chose to let it all unfold with as little intervention as possible.

He viewed his journey now as one of necessary compromise. He had not triumphed on the path he chose, and it had taken time to steady himself to the slow pace of bureaucracy. He had made bargains, but they had all been negotiated in pursuit of a goal that never changed. Longevity in this quest made him proudest. He hadn't given up. The *Green Hornet* served as a fair symbol of this adaptation: it now lived in his backyard, filled with dirt that grew herbs and vegetables. He had skirted disaster more than once in that boat, and by letting it rot, he was preventing future mistakes. Packed down with compost, the *Green Hornet* could tempt him no more.

His family. The laboratory. Those beautiful oysters. Just as passion demanded mutability, victory required tenacity. He pounded a fist into his palm and stared at me across the cafe table, adding emphasis, not anger. He spoke one lesson from his life that he could offer with certainty: "You have to persevere to do what you were meant to do."

Perseverance gave him context for interpreting the obstacles he had confronted as well as the ones still to be surmounted. As he spoke, I realized the mistake I had made in continuing to call him the Spatking years after he had hung up the mantle. With that title, I had imbued him with superpowers that he didn't care to wield. His strengths now pulsed more subtly. Here sat a sympathetic guy, known to friends and family as Justin Manley, who had learned to pace his dreams at the same speed as his abilities. Nothing fancy needed beyond his name.

Perseverance suited the story of that life just fine.

ACKNOWLEDGMENTS

A High Low Tide would not have been possible without the kindness of the following people: Justin Manley, Amelia Manley, Bryan Rackley, Tom Bliss, Charlie Phillips, Joe Maley, Danny Eller, Dan DeGuire, Dick Roberts, Karen Roberts, Earnest McIntosh Sr., Earnest McIntosh Jr., Dom Guadagnoli, Rob Hein, Randy Walker, and John Pelli. Special appreciation is due to Justin and Amelia Manley. Without their eagerness to open their lives, this book would not exist. Clearly, the lives captured in this book continue. Any necessary updates that change or finalize the story contained here will be posted on my website, www.andre-gallant.com.

    I am grateful for the multifaceted insights of Bill Walton, of the Auburn University Shellfish Laboratory on Dauphin Island, Alabama. His work along the Gulf Coast continues to enhance shellfishing in the South.

    I am also grateful to Martin O'Brien, the golfer turned oyster farmer of Cascumpec Bay Oysters in my home province of Prince Edward Island, Canada. His tour of Atlantic Canadian aquaculture operations provided me with invaluable information about the growth of the global industry.

    I gained great perspective on oystering and place by visiting the Pin Point community, located between Savannah and Skidaway Island. The Pin Point Heritage Museum is a worthy stop on any journey through Chatham County. I thank the staff of the museum for their patience. Thanks also go to Algernon Varn III, heir to the cannery, neighbor to the museum, and keeper of many salty stories, for explaining the life of an oyster factory in its waning years.

    The staff of the Georgia Historical Society Archives, at W. B. Hodgson Hall in Savannah, provided me with access to the papers of the Maggioni family as well as many other books and documents that served as a historical foundation for the modern narrative contained in this book. In addition, I am also thankful for a great many books that influenced this work of reportage: Buddy Sullivan's *Early Days on the Georgia Tidewater*, as well as our face-to-face conversation in the shadow of the St. Simons lighthouse; *Marsh Mud and Mummichogs*, by Evelyn Sherr; *The World of the Salt Marsh*, by Charles Seabrook; *Oyster: A World History*, by Drew Smith; *The Oyster Question*, by Christine Keiner; *Beautiful Swimmers*, by William W. Warner; the poetry and letters of Sidney Lanier; and the coastal writings of Nancy Willard.

I wrote and reported this book while a student in the master of fine arts program in narrative media writing in the University of Georgia's Grady College of Journalism. I toiled under the tutelage and mentorship of the program leader, Valerie Boyd, a longtime prodder and supporter of mine; Moni Basu, who gave this story focus and drive; and John T. Edge, who helped hone my voice and inspired me to expand the possibilities of my writing career. I am grateful for that trio of fine writers and educators, as well as for my colleagues who struggled alongside me to push entertaining and factual stories out into the world. To the kind editors and directors of the University of Georgia Press, I applaud your confidence in taking a risk on a young writer.

A wide community of friends and family made this project viable with spiritual and financial support over three years. I must especially thank my parents, Mary Catherine Kinney, Dwight Kinney, Paul Gallant, and Virginia Maxfield, for filling up a few gas tanks and helping with child care. Notably, without the hospitality found at the Savannah home of Coy Campbell King, I would have slept in a tent far more often than I actually did.

More than anything else, this book results from the great love of my family. To my wife, Johanna Nicol, and daughter, Mavis Nicol, know that any accomplishment in my life is in service of making you both proud.

CRUX, THE GEORGIA SERIES IN LITERARY NONFICTION

Debra Monroe, *My Unsentimental Education*
Sonja Livingston, *Ladies Night at the Dreamland*
Jericho Parms, *Lost Wax*
Priscilla Long, *Fire and Stone*
Sarah Gorham, *Alpine Apprentice*
Tracy Daugherty, *Let Us Build Us a City*
Brian Doyle, *Hoop: A Basketball Life in Ninety-Five Essays*
Michael Martone, *Brooding: Arias, Choruses, Lullabies, Follies, Dirges, and a Duet*
Andrew Menard, *Learning from Thoreau*
Dustin Parsons, *Exploded View: Essays on Fatherhood, with Diagrams*
Clinton Crockett Peters, *Pandora's Garden: Kudzu, Cockroaches, and Other Misfits of Ecology*
André Joseph Gallant, *A High Low Tide: The Revival of a Southern Oyster*